MICROSCOPY HANDBOOKS 30

Food Microscopy:
a manual of practical methods, using optical microscopy

Royal Microscopical Society MICROSCOPY HANDBOOKS

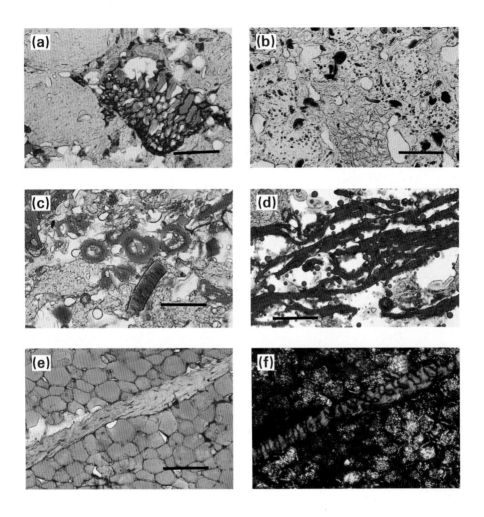

Frontispiece

Cryosections of food products. (a) and (b) 10 μm cryosections of extruded plant protein product containing soya and gluten (wheat protein). (a) Stained with toluidine blue mountant, soya protein dark blue, gluten pale blue. (b) Stained with iodine vapour, proteins yellow, degraded starch (only in gluten) purple–black. (c) and (d) 10 μm cryosections of vegetarian sausage based on soya and wheat protein. (c) Stained with toluidine blue mountant shows dark blue dough-nut-shaped particles of soya protein isolate, pink-stained palisade cells from soya bean hull and pale blue areas of gluten. (d) Stained with Oil Red O and toluidine blue mountant shows red-stained oil droplets and large fragment of textured soya protein stained purple. (e) and (f) 12 μm cryosection of bacon fat stained with Oil Red O and toluidine blue mountant. (e) Brightfield shows liquid fat in adipose tissue stained red. Connective tissue is stained lilac with purple-coloured fibroblasts. (f) Viewed between crossed polars the solid fat appears crystalline and the connective tissue shows birefringence. Figure sponsored by Bright Instrument Company Ltd.

Food Microscopy:
a manual of practical methods, using optical microscopy

Olga Flint
Procter Department of Food Science,
University of Leeds, Leeds LS2 9JT, UK

*β*IOS
SCIENTIFIC
PUBLISHERS

In association with the Royal Microscopical Society

A CIP catalogue record for this book is available from the British Library.

ISBN 1 872748 04 X

BIOS Scientific Publishers Ltd
St Thomas House, Becket Street, Oxford OX1 1SJ, UK
Tel. +44 (0) 1865 726286. Fax +44 (0) 1865 246823

DISTRIBUTORS

Australia and New Zealand
   DA Information Services
   648 Whitehorse Road, Mitcham
   Victoria 3132

Singapore and South East Asia
   Toppan Company (S) PTE Ltd
   38 Liu Fang Road, Jurong
   Singapore 2262

India
   Viva Books Private Limited
   4346/4C Ansari Road
   New Delhi 110002

USA and Canada
   Books International Inc.
   PO Box 605, Herndon, VA 22070

Dedicated to the memory of T.E. Wallis, master microscopist

Typeset by AMA Graphics Ltd, Preston, UK.
Printed by Information Press Ltd, Oxford, UK.

Contents

6. Contrast Techniques for Food Constituents   25

7. Fat in Food   35

8. Food Starches   41

9. Meat, Fish and their Products   59

10. Vegetable Proteins   79

11. The Howard Mould Count of Tomato Products   89

12. Food Gums   101

13. Food Emulsions   113

Appendix   119

Index   121

Preface

Optical microscopy can provide a rapid and effective means of examining the microstructure of food products and obtaining information which complements both chemical and physical analysis.

This manual is based on the practical microscopy courses given in the Department of Food Science at Leeds University, with additions from workshops of food microscopy for Public Analysts and scientists from the food industry.

In presenting a range of practical techniques, with the emphasis on rapid methods, the aim has been to give the reader an understanding of the mechanisms involved. After reading the manual, the microscopist should be able to select and modify techniques to make them suitable for individual food products. This approach is illustrated in the later chapters, which deal with specific types of processed foods.

Only methods which have been tried and tested by the author are included. This has meant the omission of some topics, including forensic food microscopy and food microbiology.

Olga Flint
BSc., PhD, FIFST

Acknowledgements

I would like to thank Savile Bradbury, the Series Editor, for his advice and encouragement. Thanks are also due to Adrian J. Hick who processed the films and made the black and white prints for the book and Deidre H. Williamson who produced the drawings.

Safety

Attention to safety aspects is an integral part of all laboratory procedures, and both the Health and Safety at Work Act and the COSHH regulations impose legal requirements on those persons planning or carrying out such procedures.

In this and other Handbooks every effort has been made to ensure that the recipes, formulae and practical procedures are accurate and safe. However, it remains the responsibility of the reader to ensure that the procedures which are followed are carried out in a safe manner and that all necessary COSHH requirements have been looked up and implemented. Any specific safety instructions relating to items of laboratory equipment must also be followed.

1 Introduction

1.1 The history of food microscopy

Although food ingredients and products are more widely investigated by chemical analysis, there is a long history of the microscope being used in the assessment of food quality. In 1850 Arthur Hassall showed that it was possible to distinguish between ground coffee and chicory by means of the microscope. This was at a time when the adulteration of coffee was a subject of a Parliamentary investigation. The then Chancellor of the Exchequer had said in the House of Commons 'I hold in my hand the report of the most distinguished chemists of the day who state that neither by chemistry nor in any other way can the mixture of coffee with chicory be detected'.

This statement acted as a challenge to Hassall, who bought and examined a number of samples of the coffee then being sold in London. With his microscope, Hassall identified not only chicory, but roasted wheat and beans, ground acorns, lupin seeds and powdered mangle wurzel as being present in the purchased coffees. Hassall was invited by the editor to publish his findings in the *Lancet*. These findings grew to become a series of reports on the composition of food and drugs being sold in the London shops. The reports based on over 2000 analyses not only exposed the extensive adulteration of everyday food and drugs but also gave the names of the vendors, whether the commodities were good or bad. This required Hassall to have the utmost confidence in his findings and, although he made use of the chemical tests available at the time, he relied more on the use of the microscope. The *Lancet* reports extended over the period 1851–54 and were published in book form in 1855. The findings caused immense public interest and concern which led to legislation, the first Food and Drugs Act being published in 1860. The Act sought to prevent the adulteration of foods and Public Analysts had the task of providing evidence to enforce it. In this work Public Analysts were guided by Hassall's next book *Adulterations Detected* (1857) which offered 'plain instructions for the discovery of frauds in food and medicine'. The book included advice on the choice of microscope and the simple mountants that were needed to show morphological detail, the only stain mentioned being aqueous iodine. The methods were rapid; Hassall reckoned that, by using the microscope to show tissue structure and simple chemical tests to show inorganic adulterants, 100 samples a week could be examined.

Some 50 years later, Clayton (1909) revised and enlarged that part of Hassall's second publication dealing with microscopy whilst retaining most of the original drawings. Clayton added the stains for fat and protein described by Greenish in *'Microscopical Examination of Foods and Drugs'* (1903). Here Greenish had also introduced measurement as an aid to food identification, giving the size range of granules in commercial starches and the height of palisade cells in powdered legumes. The Greenish book on 'Foods and Drugs' was quickly followed by an *'Anatomical Atlas of Vegetable Powders'* (1904) co-authored with Collin, which is still in use today.

The most comprehensive works on food structure are the four large elegantly illustrated volumes written in the 1930s by Winton and Winton. The Winton and Winton volumes are still widely consulted reference books but they give few practical details and no staining methods. For the food scientist seeking guidance on how best to use the microscope to examine food, the handbooks by Wallis had more to offer.

First published in 1923, much enlarged in 1957 and with its third edition published in 1965, *'Analytical Microscopy – its Aims and Methods in Relation to Food, Water, Spices and Drugs'* became a bible for Public Analysts and other food microscopists. Here was a laboratory manual with clear instructions for the preparation of food material for the microscope and the use of physical and chemical methods for obtaining image contrast. In the 1923 edition, the polariscope is described as 'an important accessory that will often emphasise features which would otherwise tend to be overlooked' and, in the later editions, the convenience of using discs of Polaroid is advocated.

No book dealing solely with food microscopy techniques has been published since Wallis's last edition in 1965 and since then the needs of food microscopists have changed.

1.2 Modern demands on the food microscopist

Increasingly, the microscope is being used to study the influence of ingredients and processing conditions on food structure, especially in the development of new food products. By showing the distribution and physical state of specific food constituents, particularly starches and fats, the light microscope can give a visual explanation as to why foods of similar chemical constitution have markedly different textures.

It has now become important to preserve what is often a fragile structure, especially when it is the precise nature of that structure which is of interest, as in the study of food emulsions. This makes many of the techniques used for plant and animal tissues unsuitable for food products. Histological methods have evolved to suit intact tissues which can withstand fixation, wax embedding and lengthy staining procedures. Any one of these stages may so alter a food product that the information that is subsequently obtained can be totally misleading.

It is an important principle in food microscopy that the less is done to the specimen the better. There is also a need for speed, because techniques which show changes during food processing can be used to monitor those processes. Specimens still have to be prepared for the microscope and the need for contrast in the microscope image remains, but the preparation need not be lengthy and contrast can be obtained by optical methods used alone or in combination with simple one-stage histochemical staining. The methods described in this manual have been restricted to these rapid techniques and are described in sufficient detail for the food scientist with little experience in microscopy to use with confidence. The microscope needed for this work need not be elaborate or have high power lenses but it must have suitable illumination and be comfortable in use.

References

Clayton EG. (1909) *A Compendium of Food-Microscopy.* Ballière, Tindall and Cox, London.

Greenish HG. (1903) *Microscopical Examination of Foods and Drugs.* J. and A. Churchill, London.

Greenish HG, Collin E. (1904) *Anatomical Atlas of Vegetable Powders.* J. and A. Churchill, London.

Hassall AH. (1855) *Food and its Adulterations.* Longmans, Green and Co., London.

Hassall AH. (1857) *Adulterations Detected in Food and Medicine.* Longmans, Green and Co., London.

Wallis TE. (1923) *Analytical Microscopy: its Aims and Methods.* 1st edn. Edward Arnold and Co., London. 2nd edn. (1957) J. and A. Churchill, London. 3rd edn. (1965) J. and A. Churchill, London.

Winton AL, Winton KB. (1932) *The Structure and Composition of Foods,* Vol. I. John Wiley and Sons, New York.

Winton AL, Winton KB. (1935) *The Structure and Composition of Foods,* Vol. II. John Wiley and Sons, New York.

Winton AL, Winton KB. (1935) *The Structure and Composition of Foods,* Vol. III. John Wiley and Sons, New York.

Winton AL, Winton KB. (1939) *The Structure and Composition of Foods,* Vol. IV. John Wiley and Sons, New York.

2 Choice of Equipment for Food Microscopy

2.1 The microscope

A good microscope, properly adjusted and with correct illumination, makes food microscopy rewarding and enjoyable. Recommended is a binocular microscope with 40×, 10× and 20× objectives and 10× widefield eyepieces. To this may be added a 40× objective which is necessary for work with emulsions and useful for some microbiological work. A 100× objective is unnecessary unless bacteria are to be seen in detail, the 40× lens being sufficient for viewing colonies of bacteria in food. Yeast cells, mould mycelia and mould spores can all be seen with the 20× objective.

The importance of good illumination cannot be overstressed. Preparations of food for the microscope are often thick and dense and there may be a need to view them between crossed polars which results in only a small fraction of the illuminating beam finally reaching the eyepiece. These requirements are best met if the microscope has a built-in variable intensity light source, ideally a 100 W quartz halogen lamp. Such a light source allows a polar in the form of a disc of Polaroid material to be positioned permanently just below the binocular head to act as analyser. A separate larger disc of Polaroid mounted in a photographic filter holder forms the polarizer. This can be placed over the light source, that is below the condenser, and is slipped into place and rotated to show the many birefringent features present in food, including sugar crystals, raw starch, cellulose, solid fats, collagen, etc.

The efficiency of any microscope is limited by its lighting system. This is an argument for buying a new rather than second hand microscope which may have good lenses but a poor integral light source. Before buying any microscope, test the instrument with the polars in position. Suitable readily available test subjects are smears of tomato sauce or manufactured mustard both of which contain birefringent plant tissues.

Having invested in a microscope, arrange for its annual service and ensure that users understand how to set it up to get the best image of which it is capable. The improvement that correct setting up gives is so marked and takes so little time that 'fine tuning' the instrument before starting work is always worthwhile. When help is needed, the chapter on the practical use of the microscope in the text by Bradbury (1989) is a particularly useful guide, as it provides details of setting up the instrument and also a trouble-shooting key dealing with the causes of poor image quality.

2.2   Lenses for low power microscopy

Some additional equipment is needed in addition to that required for microscopical work, the most important being some means of examining foods at magnifications below those given by the compound microscope. This link between what the naked eye sees and what the lowest power objective of the compound microscope reveals is important in the preliminary examination of many foods. It is best provided by a stereo-microscope but, if this is not available, a large (100 mm) reading glass with a magnification of 2× or 3× fixed to a stand so that the specimen may be brought into focus leaving the hands free is a useful help, although not a real substitute for a stereomicroscope which provides a better quality image and has a greater magnification range (5× – 50×).

A stereomicroscope is ideal for the rapid examination of dry food mixes, food surface faults and the cellular structure of baked and extruder-cooked products. For the latter, the field of view of the stereomicroscope (several millimetres) enables a number of air cells to be seen at the same time so that their size and shape at different sampling positions can be compared. This type of application and the robust and portable nature of the stereomicroscope make it a good 'on-line' instrument for factory use.

2.2.1 Choosing a stereomicroscope

An extensive range of stereomicroscopes is now available and most stereomicroscopes have widefield eyepieces which make for comfort and good binocular vision. Where there is a choice of eyepiece magnification, 10× eyepieces should be chosen because a higher magnification than this limits the area seen. The simplest stereomicroscopes have a magnification changer which adds a 2× and 4× magnification to the enlargement given by the eyepiece. More advanced (and costlier) models have a zoom system which gives a smooth change in magnification over a greater range, and some stereomicroscopes have additional auxiliary lenses which extend the total magnification obtainable to over 100×. The lighting for the stereomicroscope is best achieved with a separate light source which supplements or replaces any 'built-in' illumination. For work with food surfaces a modern fibre optic light source powered by a quartz halogen lamp is ideal. This provides an intense but cold light which does not dry the specimen and the single, or preferably double, swan necks which carry the bundles of light-transmitting fibres can be angled to give the optimum contrast to the surface detail of specimens.

Some stereomicroscopes are equipped with a transmitted light source which can be useful, particularly when combined with incident lighting, for viewing thick preparations of translucent subjects such as cake or bread crumb.

2.3   Accessories

A small amount of additional equipment is needed for microscopical work with food; the following list includes the most useful items.

- Scissors (5 in)
- Forceps: large blunt; fine-pointed; watchmakers curved
- Small fine-pointed forceps
- Micro-spatulas or section lifters
- Single-edged cutting needles
- Single-edged stainless steel razor blades
- Squirrel hair paintbrush
- Simple droppers with rubber teats
- 25-ml and 50-ml dropper bottles for stains and reagents
- Embryo dishes (hollowed glass blocks) with internal diameter 45 mm
- Glass Petri dishes (74 mm and 120 mm)
- Sieve (5 in/128 mm) diameter 180 μm aperture
- Set of cork borers (for sampling)
- Stainless steel mesh tea strainer or coarse sieve
- Straight-edged kitchen knife
- Plastic chopping board

For measuring work with the compound microscope, a widefield focusing eyepiece fitted with an eyepiece graticule and a 1- or 2-mm stage micrometer will be needed. For measuring with the stereomicroscope a short (150 mm) metal ruler calibrated in 0.5-mm divisions which can be viewed at the same time as the specimen will be found useful. Accessories for use with the cryostat are listed later.

Reference

Bradbury S. (1989) *An Introduction to the Optical Microscope,* revised edn. Oxford University Press, Oxford, pp. 70–75.

3 Preparation of Food for the Stereomicroscope

3.1 Introduction

The nature of the food sample and the information required about it largely determine the specimen preparation needed but, before preparing material for the compound microscope, it is often worthwhile discovering what information the stereomicroscope can yield.

Little time is needed to prepare material for the stereomicroscope and, as well as providing information on the nature of the specimen, observation with the stereomicroscope can be a guide to the preparation that is needed for work with the compound microscope. Powders and other particulate foods such as dry soup and cake mixes should always be examined with the stereomicroscope because this alerts one to the presence of minor constituents which could be missed when viewing the small amount of material used to prepare a slide for the compound microscope.

3.2 Preparation of powders

Thoroughly mix the sample and weigh out 1 g, place in a metal mesh tea strainer or coarse sieve and sift this over the base of a large (120 mm) Petri dish; this provides the fine fraction. Any coarse material retained by the strainer is placed in another Petri dish to be examined separately.

Ensure that each fraction is spread evenly over the base of the Petri dish by tapping or jarring the dish if necessary. Place on a black background and examine both fractions with incident light using the whole magnification range of the stereomicroscope. Make a note of the different constituents present and, using fine curved watchmakers forceps, pick out any minor constituent which has characteristic features, for example particles of spray-dried colour, placing these in a hollow glass block for later examination. Next, tip each dish slightly and vibrate it by jarring.

This causes sugar and salt crystals to separate from starch and other powdered constituents, and enables them to be picked out easily.

The fine fraction can be mounted and studied using the compound microscope, for example to determine the type of starch present and, if necessary, further fractionation can be achieved by using the 180 μm sieve. The coarse fraction from the sieve needs further preparation to make it suitable for the compound microscope.

3.3   Preparing surfaces

The stereomicroscope is ideal for the examination of surface defects of food products, including the presence of mould hyphae, sugar, salt and fat deposits which are responsible for unsightly patches or lack of surface gloss on food. A variation in surface colour may be due to a deposit or to a feature lying just below the surface, for example trapped air bubbles or the presence of a poorly dispersed ingredient.

Surfaces require little preparation but they must be well lit. Illuminate the surface from the side, ideally with a fibre optic light source, adjusting the angle of the light beam so that it just grazes the surface. This will show whether the fault lies on or just below the surface. A surface growth or deposit can be removed using Sellotape's 'Invisible deluxe tape'. This is a translucent sticky tape which is similar to 'Magic' mending tape but is not birefringent. Before removing a surface deposit, put a drop of light machine oil (e.g. 3-in-1 oil) on a microscope slide, collect the deposit with a short length of tape and lay the tape on the oil drop with the deposit uppermost. Add one or two drops of oil to cover the deposit and carefully position a coverslip over it. Examine the slide with the compound microscope using the 4× and 10× objectives in brightfield and with the polars partly and fully crossed. Sugar and solid fats are crystalline and birefringent, salt (sodium chloride) is crystalline but not birefringent. To distinguish sugar from crystalline fat, warm the slide very gently; any fat will melt and dissolve in the oil whereas sugar is unaffected. Examine the slide whilst it is still warm, as fat crystals may reappear as the oil cools.

3.3.1 Sub-surface faults

Some surface defects are due to faults that lie just below the surface. A good view of these can be obtained by 'clearing' the surface layers. This is done by putting a few drops of light machine oil on the area of the fault, this reduces light scattering by the surface and allows some light to penetrate to sub-surface features before being reflected by them. The result is a darker coloured surface which has a translucent layer in which any abnormal features can be seen more clearly.

3.4   Preparing internal structures

For all products, the prime requirement is a really flat, level surface that can be positioned parallel to the objective of the stereomicroscope. This is achieved in different ways depending on the moisture content of the specimen.

Dry brittle products such as biscuits, crackers and especially extruder-cooked snack foods are first broken or cut to reveal an internal surface which is then trimmed with a razor blade or sandpapered with a medium and fine sandpaper to give a plane surface. Any dust adhering is removed with a fine squirrel-hair paintbrush. The specimen is then fixed to a microscope slide, preferably a large slide (38 × 75 mm), using 'Blu-Tak™' or plasticine and positioning it with the prepared surface in a horizontal plane so that it will be parallel to the objective of the stereomicroscope. To obtain the most information on the structure, lighting, focusing and magnification are all important. If a double swan neck fibre optic light source is available use one light source close to the specimen at a grazing angle and the other further away and at a steeper angle. This is particularly important if the specimen is to be photographed. If a zoom stereomicroscope is used, first focus at the highest magnification, then return to the lowest magnification and zoom up to the optimum enlargement. This will ensure good focusing throughout the zoom range.

With care, a succession of surfaces can be exposed through products with a biscuit-like shape but, for most purposes, the preparation of two surfaces cut at right angles to one another will suffice. Such longitudinal and transverse surfaces (analogous to the longitudinal and transverse sections used by biologists) provide useful information for extruder-cooked products, particularly those which have an aerated structure. The aeration usually takes the form of air pockets elongated in the direction of extrusion. With some products, especially those based on soya protein, the distension is so marked that the longitudinal surface appears to show a fibrous structure (see Chapter 10, *Figure 10.4*).

Some moist foods (e.g. sliced bread, cut cake and sliced meat products) need little preparation because they already possess a flat surface parallel to the base of the slice. All that is needed is that the slices are sampled to show features of interest, and for this a large diameter cork borer is ideal. Such mechanically produced slices provide excellent subjects and illustrate the advantage of having the surface to be examined exactly parallel to its base. This should be the aim when unsliced products are prepared for examination.

Unsliced moist products include canned meats, chunks from canned and frozen vegetarian meals and jellied products, as well as many moist baked items. The slices which need to have parallel upper and lower surfaces can be cut with a straight-edged sharp knife, a section razor or a single-edged razor blade. It is easier to cut several small slices from different areas rather than to attempt sizes similar to those of ready-sliced foods. The examination of the prepared sample is similar to that described under dry products.

3.5 Staining preparations

Although most specimens for the stereomicroscope rely on lighting for image contrast, a limited amount of staining is possible. Most of the staining methods described later can be adapted for surface staining. Iodine staining is particularly useful because it shows both protein and starch. Aqueous iodine can be used on meat products and iodine vapour can be used directly for semi-moist items such as bread and cake. Dry products need to take up some moisture before iodine vapour staining. This can be done by leaving the specimen over water in a closed container for 1–3 h before staining, the exact time depending on the nature of the product.

4 Simple Specimen Preparation Techniques for the Compound Microscope

4.1 Introduction

For examination by the compound microscope, a specimen needs to be translucent, that is thin enough to transmit light when mounted in a suitable medium. Translucency is achieved in ways that depend on the physical nature of the food, for example powders are sufficiently thin to be made into wholemounts, and viscous foods like sauces can be prepared as smears or squashes. Solid foods, especially meat products, are best prepared as thin sections by using a cryostat, although some work can be done with smears of these (see Section 9.8). Preparing specimens for the microscope is such an important part of food microscopy that each of these techniques is described in detail.

4.2 Wholemounts

Wholemounts consist of particles usually less than 200 µm in thickness mounted in a medium which allows the particles to be seen without dissolving or unduly swelling them. The mountant may contain a stain which can help in the identification of the material.

A great many particulate foods are suitable for wholemount preparation, including starch granules, cereal and legume flours, dried potato and onion, spray-dried milk, coffee and encapsulated flavours and colours, all the hydrocolloid powders (guar, pectin, carrageenan, etc.) and crystalline materials such as salt, sugar and monosodium glutamate. All these can be studied by mounting in an inert oil and/or a solution of a suitable stain.

If the powdered material appears at all heterogeneous in particle size, sift the product into two fractions using the 180 µm sieve. The fine fraction provides the material most suitable for direct mounting; the coarser fraction, which to some extent will have been cleaned by the sieving, may require squashing or sectioning to make it suitable for examination. Often all the information that is required is found in the fine fraction because this will usually include fragments of the larger particles.

Start by mounting the powder in an inert oil such as a light machine oil (e.g. 3-in-1 oil). This enables the powder to be seen without swelling or loss of water-soluble constituents. The oil is also a good medium for observing the birefringence of raw starch, sugars and the lignified cellulose of plant constituents. Prepare the mount slowly and carefully, as this avoids the inconvenience of air bubbles trapped in the preparation. Put a small drop of mountant on the slide; take a little powder on the edge of a micro-spatula and tap the powder on to the mountant. Mix with a needle and allow any obvious air bubbles to escape. Apply a coverslip by resting one edge of it on the slide alongside the specimen and carefully lower the slip with a needle, allowing the mountant to fill the space between the slip and the slide, thus avoiding trapped air bubbles. Remove excess mountant by inverting the slide and pressing very gently on fluff-free blotting paper.

4.3   Smears

Smears are made from viscous foods containing fine particles or droplets which can be spread to give a translucent layer, for example sweet or savoury sauces, fruit or vegetable purées and emulsions such as salad creams.

Place a small amount of sample towards one end of a microscope slide resting on the bench. Use a second slide inclined at an angle of 45° to draw the sample gently across the slide to give an even and translucent layer. Add either 1 drop of water or aqueous stain and cover as described for wholemounts. For emulsions which break easily, for example modern very low-fat spreads, use a coverslip to spread the product by gently easing a small blob across the slide and examine uncovered as well as with a coverslip. Avoid using pressure when the coverslip is applied.

4.4   Squashes

The coarser fractions obtained by sieving and other material of sizes greater than 180 μm may not smear well even when fully hydrated. Apart from sectioning, squashing is the only way by which such material may be made translucent; this has the merit of providing quick results, but structural features are often distorted.

Squashes are useful for showing raw starch in baked and extruded products such as sausage rusk and breakfast cereals. Yeast cells can be shown in bread crumb and mould mycelia in cheese. Cereal bran, and the hull and cotyledon tissues from soya beans and peanuts can be identified easily from squashes.

4.4.1 Method

Satisfactory squashes are best prepared from fully hydrated samples. Even when the food product appears to be moist, squashing will be easier and less damaging if the sample is first soaked in water. If sufficient material is available cut into two or more samples, place these with water in a hollow glass block or Petri dish and cover. Leave for a few minutes, longer if the material has not softened. If the material is very dark coloured, put some of it to clear in diluted household bleach (1 part bleach to 5 of water) and leave until lighter in colour (15–20 minutes), then transfer the sample to clean water before squashing.

4.4.2 Squashing technique

1. Place a small amount of the moist sample on one slide and place over and at right angles to it a second slide.
2. Press the slides together and shear them with a twisting movement until the sample appears translucent.
3. Part the slides and add a drop of water or aqueous stain (e.g. iodine or toluidine blue) to each.
4. Position a coverslip over the sample, invert and blot on fluff-free blotting paper using moderate pressure.
5. Examine with 4× and 10× objectives, with brightfield illumination and between crossed polars.

This method is obviously very disruptive, but it shows the presence of constituents such as starch or fat, especially birefringent starch (i.e. raw starch), birefringent fat (solid fat) and oil globules (liquid fat).

Where the evaluation of structure and the relative position of constituents is of interest, cutting frozen sections with a cryostat is much to be preferred.

5 Use of the Cryostat in Food Microscopy

5.1 Introduction

For most manufactured foods, the demonstration of structure and the distribution of constituents requires sections. Sections can be of wax- or resin-embedded material, but a more rapid method of obtaining sections, and one which preserves fat and causes the least change in the material, is by the use of a cryostat. A cryostat, which consists of a microtome housed in a freezer cabinet, enables sections to be cut from food products by rapidly freezing the water naturally present in, or added to, the food, the resulting ice acting as an embedding medium.

A major advantage of using a cryostat to prepare food sections is the speed with which results can be obtained. Sections can be cut and stained within minutes. Constituents, including fats, are retained in their original sites and there is no need for fixation, although fixed material may be sectioned.

Cryostat sectioning is the best method for preparing what may be termed fabricated foods, that is all those foods which owe their structure to the method of manufacture, rather than to nature. These include baked and extruded products, and most importantly, meat products, regardless of the processing these have received. Both fine and coarse particles can be sectioned with the cryostat, including those previously fractionated by sieving. Spray-dried products, including spray-dried vitamins, flavours and colours, are also suitable for cryostat sectioning. One advantage of sectioning particles is that the particles are seen as they would appear in a sectioned foodstuff. Samples of protein particles such as soya protein isolate, soya grits, wheat protein powder (gluten) and also herbs and spices can be sectioned to provide reference slides to assist in the identification of these as they occur in cryosections of comminuted meat products.

5.2   Choosing a cryostat

A modern open-top cryostat (see *Figure 5.1*) is easier to work with than the earlier 'port-hole' type models, although these too can yield satisfactory sections provided that the microtome has a retracting mechanism. This is necessary for sectioning food products because retraction of the specimen holder on its upward movement prevents the specimen pressing on the back of the knife which can cause smearing, or loss of constituents. This situation can arise easily if ice has collected on the back of the knife.

Most cryostats can maintain an internal temperature in the range of –20 to –30°C, which is suitable for almost every food including some which are rich in fat, such as sausages and nut butters. For extremely fatty foods like margarine, or those with a high sugar content, such as cake batters, a cryostat needs to maintain a temperature of about –38°C which is best provided by a machine designed to operate down to –45°C.

A new cryostat will be supplied with a knife which has either safety cut-outs at either end or knife guards which serve the same purpose. At least one, and preferably two, spare knives will be found useful for when the first knife is being sharpened. Also supplied or offered will be specimen holders of different sizes. The most convenient type of specimen holders are hollow (i.e. tubular) and have two opposing 5-mm holes; tubular holders cool more quickly than solid ones and the presence of holes makes the cooled holder easy to grip with blunt forceps. Lubricants and an artist's long-handled bristle brush used for knife cleaning complete the accessories normally supplied with the cryostat.

Although modern cryostats usually have an automatic defrost cycle, they can be kept entirely free from ice by placing several net bags filled with self-indicating silica gel in the microtome compartment. These bags are easily regenerated by oven heat.

5.2.1   Quenching

An important accessory for cryostat work is some means of efficient quenching, that is the rapid freezing of the specimen prior to sectioning. Liquid nitrogen or a bench carbon dioxide Fast Freezer provide the best all round coolants for the wide variety of material that a food microscopist may need to handle, but some users recommend Cryospray, an aerosol spray of dichlorodifluoromethane. Liquid nitrogen is perhaps the most versatile of these quenching agents.

If liquid nitrogen can be used, a small (500 ml) Dewar flask for quenching and a larger (2 or 5 l) Dewar flask or insulated container for nitrogen storage will be required. Also needed for quenching is a long-handled holder with a metal mesh sleeve or pocket into which the specimen holder fits loosely, but securely.

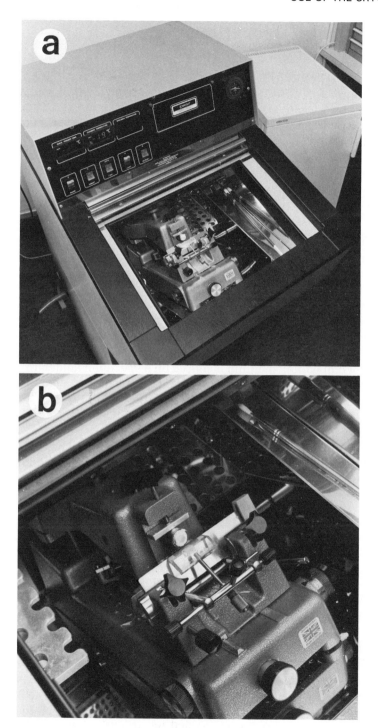

Figure 5.1: (a) An open-top cryostat. (b) Interior of cryostat showing guide plate in position on microtome blade.

5.3   Pre-cryostat operations

Although the method of quenching can be the same for all specimens, the method of sampling and the preparation needed before quenching varies with the nature of the food specimen which may be moist or dry, dense or aerated, and in an entire or powdered state.

5.3.1   Preparation and quenching of moist dense products

Moist foods that are not aerated and which can be cut easily with a knife, for example canned meat products and cooked sausages, are cut into 3–5 mm thick slices which are then sampled using a no. 6 cork borer (diameter 10 mm). This gives disc-shaped specimens with a flat surface suitable for mounting on a specimen holder. Pasty products such as raw sausage meat and meat pâtés are first formed into small balls which are gently pressed to give disc-shaped specimens 3–5 mm thick with a diameter of 5–10 mm.

Next, a drop of embedding medium is placed on a specimen holder, the specimen is positioned on it and covered with more embedding medium. The specimen is now ready for immediate quenching.

Quenching with liquid nitrogen.
Safety note: as liquid nitrogen boils at around –196°C, it must be handled with great care at all times. Gloves must be worn when pouring liquid nitrogen, and non-spectacle wearers must wear safety spectacles throughout its use.

1. Have ready a small (500 ml), wide-mouthed Dewar flask two-thirds full of liquid nitrogen.
2. Place the specimen holder with its sample in a long-handled mesh carrier, and gently lower the carrier into the liquid nitrogen until half of the specimen holder is submerged. Wait until the embedding medium begins to whiten, which indicates the start of freezing, and then fully immerse the specimen; leave submerged until the turbulent boiling of the liquid nitrogen ceases.
3. Remove the carrier and transfer the specimen holder to the cryostat. This can be done easily and safely if the specimen holder is a tubular one with holes in opposite sides so that it can be gripped firmly with large blunt-ended forceps.
4. Leave the specimen in the cryostat for 20 min to allow it to reach the temperature of the cryostat. For many food samples this will be set at –20°C but, for samples with a high fat or sugar content, a lower temperature (–25 to –35°C) may be needed. The block is now ready for

trimming and sectioning (see Section 5.4). Blocks should not be left in the cryostat much longer than about 2 h as they tend to dehydrate, become brittle, and are then difficult to section.

5.3.2 Preparation for quenching of moist aerated products

Foods that can be sliced easily but which contain air spaces, for example bread and cake, need to have most of the air replaced by an embedding medium before quenching, otherwise they will be difficult to section.

1. Prepare 3–5 mm thick slices and sample these with a no. 6 cork borer.
2. Place the resulting discs in hollow glass blocks containing embedding medium. Turn the discs over using a spatula or section lifter and press and release the discs very gently to expel most of the air so that the embedding medium takes its place.
3. Gently lift the impregnated disc on to a specimen holder containing a drop of embedding medium and cover with more medium.
4. Quench immediately as described in Section 5.3.1.

5.3.3 Sampling and preparation for quenching of dry foods

Dry brittle foods which tend to shatter when cut, including many extruder-cooked items like texturized soya protein (TSP) and starch-based snack foods, are easier to handle if they are first softened. TSP and other high protein extrudates can be soaked in water but any starch-based extrudate is best softened by storing in a moist atmosphere, for example in a small desiccator with its base filled with water. The time needed for softening varies from less than an hour to several hours depending on the ambient temperature and the product, which may swell or even collapse if left too long in a warm moist atmosphere. When the sample is just soft enough to cut, prepare 3–5 mm slices noting their orientation, for example cutting both longitudinal and transverse slices if the extruded product has a cylindrical shape. Use a cork borer to cut specimens from the slices if these are large enough. Treat with embedding medium and prepare for quenching as in Section 5.3.1 or 5.3.2 according to whether the product is aerated or not.

5.3.4 Preparation of spray-dried powders

Because of their method of manufacture, spray-dried powders, such as coffee, milk, soya protein isolate and encapsulated vitamins and flavours, disperse rapidly in any aqueous medium and so preparing them for quenching must be done very quickly. Success with such powder depends upon having to hand before starting all the necessary equipment and materials, including fridge-cooled embedding medium and liquid nitrogen in the quenching flask.

1. Stir a few milligrams of the powder into the cooled embedding medium contained in a hollow glass block.
2. Using a small spatula, quickly scoop up one or two drops of the suspension and drop these on to a specimen holder.
3. Quench immediately as described in Section 5.3.1.

5.3.5 Food particles and powders other than spray-dried products

Milled and drum-dried products are less sensitive to moisture than spray-dried products and may benefit from hydration in an embedding medium. These include soya grits, dried gluten (wheat protein) and the coarse fractions (180 μm) derived from sieving.

1. Stir a little of the particulate material into the embedding medium contained in a hollow glass block.
2. Cover and leave for a few minutes to hydrate and to allow any trapped air to escape.
3. Stir and scoop up the suspension with a small spatula and drop on to the specimen holder.
4. Quench promptly before the drops of medium run out, resulting in a specimen that is too thin for sectioning.

5.4   Cryostat sectioning with a rotary microtome

1. Clamp the specimen holder in position and move the knife carriage until it just touches the specimen block, and firmly clamp the carriage.
2. Trim the block by cutting thick (30 μm) sections until the embedded material is just visible.
3. Select the section thickness required, the intention being to cut sections one feature thick. A thickness of 10 μm is a good starting point for new material and, since cryosections are of hydrated, unfixed material, a 10-μm section is equivalent to a much thinner wax section. Cut several sections with the guide plate clear of the knife, observing the quality of the sections and the ease of cutting. If the sections crumble or compress this is most likely due to a blunt part of the knife, so move the knife along to give a fresh cutting edge.
4. When the block appears to be sectioning well, clear any accumulated debris from the knife with the long-handled brush, which should be

kept in the cryostat. Brush away from the knife base towards the cutting edge to prevent damage to the knife and the bristles of the brush.

5. Position the guide plate so that it rests on the knife with its top edge parallel to the knife facet and a tiny bit above it. If the guide plate is set too high the block will collide with it, too low and the section will start to curl and not pass between the guide plate and the knife. Positioning the guide plate is a crucial stage in obtaining good sections. It is best to make adjustments slowly and carefully following the manufacturer's instructions.

6. With the guide plate in position and a supply of microscope slides to hand, cut the section to be collected which should lie on the knife blade. Then lift the guide plate very gently (so as not to create an air current which could dislodge the section) and collect the section by taking a clean microscope slide at ambient temperature and holding it near to, and parallel to, the section. Do not press the slide on to the section, which is attracted to the warmer slide and will move on to it if the slide is held near to, but not touching the section.

7. Stain the section either with the toluidine blue stain mountant (Section 6.3.1) or by iodine vapour staining (Section 6.3.2) for sections of starch-based products or any delicate or water-dispersible sections such as those of spray-dried products. Examine the stained section to see if it appears too thick (i.e. shows overlapping features) or, which is more likely, it is too thin and shows incomplete features (e.g. an aerated product might show only fragments of its cellular structure). Adjust the section thickness as appropriate and cut, stain and examine further sections until the optimum section thickness has been established. In some cases it is useful to have both thin and thicker sections as these show different features. Sections a little on the thin side yield better photomicrographs with higher power objectives, whereas well-stained thicker sections are appropriate for the 10× and lower power objectives including those of the stereomicroscope.

8. Collect sections of optimum thickness and store the slides on well-labelled cardboard slide trays which can be kept in the refrigerator. Such stored slides will keep satisfactorily for some weeks. Reference slides which need to be kept longer can be stored in the deep freeze at –20°C for several months. The most obvious change that occurs on prolonged storage is for neutral fats to become acidic so that they colour with basic dyes such as toluidine blue. There is also a tendency for unmounted sections to dehydrate, so much so that they need a period in a moist atmosphere to restore their ability to take up iodine vapour. Rehydration can be done quickly by using a large Petri dish as a moisture chamber. Place a filter paper in the base of the dish. Saturate with water and put a single microscope slide on the wet paper. Position the section-bearing slide with section uppermost on the empty slide and replace the Petri dish lid. Leave for 10 min or longer. The slide is now ready for staining.

6 Contrast Techniques for Food Constituents

6.1 Introduction

Once a food material is in a state suitable for microscopical examination, that is, thin enough to have the potential to transmit light, the next stage is to choose a suitable mountant. Left in air, and viewed by transmitted light, the difference between the refractive index of a dry foodstuff (n = approximately 1.5) and air (n = 1.00) is so great that light rays are unable to pass through even a thin preparation of food which then appears dark and featureless. A mountant enables light to pass through the specimen and this allows detail within the specimen to be seen. Smears and squashes already contain water which acts as a mountant, and aqueous staining solutions also behave as mountants. A disadvantage of such aqueous mounts is that they cause the loss of water-soluble constituents such as sugar, salt and spray-dried material. So, for some specimens, especially mixed powders of unknown constitution, a preliminary study using an inert mountant and physical (i.e. optical) methods of obtaining contrast, especially the use of polarizing filters (polars), is worthwhile.

6.2 Physical methods of obtaining contrast

6.2.1 Choice of mountant

A mountant needs to have a refractive index different from that of the specimen otherwise the specimen will be invisible unless it happens to be coloured or is birefringent and is viewed between crossed polars. A mountant of lower refractive index than the food is preferable, because the outline shadows in the image are then around the features present in the specimen rather than within them.

Oils as mountants. An inert oil such as any light machine oil (e.g. 3-in-1 oil) is the safest mountant if water-soluble material is not to be missed.

Alternatives are liquid paraffin or an edible oil such as soya or rapeseed oil. Both powders and sections can be mounted in oil as described for wholemounts.

Examine the oil mount by brightfield illumination, and between partially and fully crossed polars using the 10× and 20× objectives.

Brightfield illumination will show the shape of particles, enabling spray-dried powders to be distinguished from drum-dried products. When examined between crossed polars, sugars, raw starches, lignified structures, for example bran and also crystalline material (other than sodium chloride), all show birefringence. Partly crossed polars enable the birefringent material to be seen in relation to isotropic structures (e.g. salt) because, although crystalline, salt is not birefringent.

Water as a mountant. Because of its low refractive index (n = 1.33), water gives a mount showing the greatest contrast in the microscopic image. An aqueous mount is especially useful for showing small amounts of raw starch in sections or powdered food mixes, especially if the slide can be viewed with partly crossed polars. An aqueous mount can also be used to show the physical properties of the mounted specimen, for example its ability to swell or dissolve in water. This may be compared with its appearance when mounted in oil.

With care, both an oil and a water mount can be made on the same slide and this makes the comparison easier. The dissolving of material in the aqueous mount which appears unaffected in the oil mount is the first difference to be seen. Then, by warming the slide very gently, and to no more than about 40°C, any solid fat present will be seen to change. At temperatures above 37°C fats melt and begin to dissolve in the oil mount but form oily globules in the water mount. On cooling, the globules will be seen to contain crystalline fat which is visible when the slide is viewed between partly crossed polars.

These physical ways of obtaining information can be used to suggest confirmatory staining methods which are a means of gaining contrast dependent on the chemical nature of the material.

6.3   Chemical methods of obtaining contrast

Complementing the optical methods of obtaining contrast are staining techniques that involve known chemical mechanisms. These histochemical methods can be relied on to demonstrate the main groups of food constituents: carbohydrates, proteins and fats, unequivocally and, within each food group, important members can also be identified.

The periodic acid–Schiff reaction for carbohydrate (Hotchkiss, 1948) the hydroxylamine–ferric chloride method for pectin (Jensen, 1962) and the picro-Sirius Red technique for collagen (Flint and Pickering, 1984) are all examples of histochemical methods that have been applied to foods. These methods have the advantage of yielding permanently stained slides, but they are all lengthy multistage techniques. For most purposes, the less permanent but rapid single-step methods which avoid loss of section constituents are to be preferred. A good example is the use of toluidine blue as a stain mountant.

6.3.1 Toluidine blue as a general differential stain

Toluidine blue (CI 52040) is a basic dye widely used for staining because it has metachromatic properties which can be used to identify the many food constituents which contain anionic groups. Identification depends on the fact that different food constituents have different dye-binding properties and this is reflected in the ways that the bound toluidine blue molecules are aligned. Where the charged anionic groups are close together (0.45 nm or closer) the dye molecules are aligned so that dye polymers are formed, and these have a distinct red (magenta) colour, which contrasts well with the bound monomer, which is blue.

Accessibility also plays a part, which may explain why some constituents are coloured turquoise or blue–green (e.g. elastin and lignified plant structures).

Toluidine blue can be used in the form of an aqueous stain mountant (Flint and Firth, 1988). The mountant has a dye concentration which, if it were in water alone, would have a strong purple colour showing the presence of toluidine blue polymers, but the dye solution contains glycerol and phenol which together restrict polymerization so that the dye–mountant is blue in colour showing that the toluidine blue is largely present in a monomeric form. Polymerization occurs 'in situ' when the dye–mountant is put on a section and, because the dye is concentrated, a differentially stained preparation is produced within 2 min with 10-μm cryosections.

Choice of toluidine blue. It is important to use a dye sample that is known to contain at least 85% dye. Some commercial samples of toluidine blue consist of the toluidine blue–zinc chloride double salt and these contain only 50–58% of dye; such samples are unsatisfactory, even when used in correspondingly large amounts (Firth and Flint, 1988). Toluidine blue certified by the American Biological Stain Commission is recommended. Although the Commission specification is for a minimum of 50% dye, most of the samples approved since 1983 have a dye content in the range of 85–94%. These are available from both Sigma and Aldrich.

Preparation of a toluidine blue stain mountant. The staining medium consists of a 0.049% solution of toluidine blue (based on dye content) in 30% aqueous glycerol and contains 1.01% phenol.

Dissolve sufficient toluidine blue in distilled water to give the equivalent of 70 mg of pure dye per 100 ml. Add 30 ml (37.8 g) of glycerol to each

70 ml of aqueous toluidine blue, mix and add exactly 1.01 g of phenol to each 100 ml of stain mountant. Allow to stand overnight.

Using a dropper place one or two drops of the staining medium on the section, smear or squash so that it is completely covered. Leave for 1 min, apply a coverglass and leave for another minute before inverting and pressing very gently on blotting paper to remove excess stain. The slide is then ready for examination.

The results seen with brightfield illumination are summarized in *Table 6.1*.

Table 6.1: Colours of food constituents stained with the toluidine blue stain mountant

Muscle fibres	Pale blue with fresh meat products more purple if phosphate is present, darker blue with heat-treated products
Muscle fibre nuclei	Red–violet
Raw collagen	Pale pink
Cooked collagen	Pale lilac
Fibroblasts	Blue–violet
Elastin fibres	Turquoise
Soya protein (grits, isolate, textured soya protein)	Dark purple–blue
Soya and other plant cell walls	Magenta
Wheat protein (gluten)	Pale blue–green
Lignified cellulose (present in onion, bran, spices, etc.)	Dark blue or blue–green
Rusk (except protein)	Unstained
Fat	Unstained (unless acidic)
Fatty acids	Pale blue
Food gum particles (except starch-based gums)	Pink, purple, magenta. See Chapter 12 for identification of individual gums

Toluidine blue staining of acidic foods. Foods that contain acetic or other organic acids have a reduced capacity to stain with toluidine blue because the anionic groups which bind the dye are not fully charged at low pH values. This makes the toluidine blue mountant less effective with tomato ketchup and other vinegar-based sauces. For these and other acidic products a simple 0.05% aqueous toluidine blue gives better results enabling herbs, spices and incompletely dispersed thickeners, such as guar and xanthan gum, to be seen.

Method for food smears containing vinegar. Add 1–2 drops of 0.05% aqueous toluidine blue to the smear, leave for 2 min, apply a coverslip, blot and examine in brightfield and with crossed and partly crossed polars. The results are similar to those given by the stain mountant but with less colour contrast between different constituents.

6.3.2 Iodine as a histochemical reagent

The use of iodine is regarded as the classic method for demonstrating starch, but iodine is more than a simple test for starch. At an appropriate strength, iodine can be used to demonstrate proteins and it is also a useful test for several food gums (see Chapter 12). Iodine has only a limited solubility in water and an iodine stain is usually made by dissolving iodine in aqueous potassium iodide but it may also be made by diluting with water the alcoholic solution of iodine sold by pharmacists as Tincture of Iodine BP. In the presence of moisture, iodine vapour is a very useful stain. The vapour staining of starch is appropriate for the wide range of foodstuffs containing heat-processed (gelatinized) starch which tends to be swollen by aqueous iodine.

Iodine's full potential is realized when the optimum concentration of iodine and the form in which this is used (i.e. aqueous or vapour) to give strong or weak staining is chosen to suit the material being examined. For example, a 1% aqueous solution of iodine gives starch an immediate blue–black colour which is useful for detecting the presence of starch. However, such a dense colour masks any birefringence the starch may have; a very dilute solution of iodine on the other hand hardly colours intact or chemically cross-linked starches but is ideal for staining gelatinized starch and allows the birefringence of intact granules to be clearly seen.

Different concentrations of iodine are best made by diluting a stock solution of iodine just before use. Weak solutions of iodine do not keep well because iodine is easily lost by exposure to light and is readily absorbed by the rubbery materials used for teats in staining dropper bottles.

Preparation of stock iodine solution.
 1 g iodine
 2 g potassium iodide
 100 ml distilled water
First dissolve the potassium iodide in the water then add the iodine and stir until dissolved. Store in a glass-stoppered amber coloured glass bottle.

A working solution of iodine is prepared by diluting 5 ml of stock solution to 100 ml with distilled water, just before use.

Method for aqueous iodine staining. For cryosections, cover the section with 1–2 drops of freshly diluted iodine solution. Leave for 1 min, add a coverglass, invert and blot gently on fluff-free blotting paper to remove excess stain.

Examine the slide by brightfield illumination and between crossed polars using 10× and 20× objectives.

For powders, sprinkle 2–3 mg of the powder on to a microscope slide. Tap the slide to give an even layer of the powder about 10 mm in diameter, add 1–2 drops of freshly diluted iodine solution and proceed as for cryosections (see above). The results are summarized in *Table 6.2.*

Table 6.2: Aqueous iodine-stained starches, dextrins and proteins viewed brightfield and between crossed polars

Raw, amylose-containing starches	Blue, birefringent
Raw, amylopectin starches	Reddish, birefringent
Chemically modified starches	Yellow or brown, birefringent
Pre-gelatinized, i.e. precooked starch powders	Reddish, swollen particles, which may contain individual blue-stained granules (not birefringent)
Dextrins	Blue–purple or reddish particles which quickly disperse
Proteins in cryosections and powders	Yellow

Iodine vapour staining. The use of aqueous iodine can be a fairly satisfactory method for showing the presence of starch quickly but it has certain drawbacks. The aqueous solution may cause swelling of gelatinized starch and the dispersion of any loosely held starch. The use of iodine vapour avoids these problems and also makes it easier to control the degree of staining. This makes it particularly suitable for showing gelatinized starch in food emulsions, and baked and extruder-cooked cereal products, enabling starch to be seen correctly located in relation to other constituents (Flint, 1982).

Safety note: iodine vapour is poisonous and, ideally, iodine vapour staining should be done in a fume cupboard. If a fume cupboard is not available a glass desiccator can be adapted to take the vapour-staining chamber by filling the base of the desiccator with a saturated aqueous solution of sodium thiosulphate, and placing the staining chamber on a perforated tray above this. The vapour-staining chamber should remain in the fume cupboard or the desiccator at all times.

To prepare the staining chamber, cut a piece of fluff-free blotting paper (Postlip) the same width as a rectangular glass staining jar and 20 mm longer than the base of the jar. Fold the ends to make a paper tray and place on the bottom of the jar so that 10 mm of paper fits against each of the end walls. Moisten the paper and sprinkle on it a thin layer of iodine crystals. Immediately replace the lid of the staining jar and place in the fume cupboard or prepared desiccator (see safety note). Leave for about 10 min to allow the iodine vapour to reach equilibrium. The vapour chamber is now ready for use.

1. For optimum staining, the specimen requires a film of moisture. To achieve this, cryosections may be left in the cryostat or deep freeze cabinet (−20°C) for 1 min and then brought to room temperature or, alternatively, simply breathe heavily on the specimen; this is all freshly sectioned cryostat sections need. Smears of food emulsions may not need this hydration step but some benefit by moist breath. Treatment at −20°C as an aid to hydration is not recommended for food emulsion smears.

2. Next place the slide horizontally in the iodine vapour chamber resting it on the ends of the paper tray with the section or smear facing down. Quickly replace the staining jar lid.

3. Observe the progress of staining through the glass of the container and when some darkening of the specimen has occurred, which may be in as little as 30 sec, take the slide out and examine it (without mounting) using the 10× objective. Take care that the unmounted specimen does not touch the objective.

4. If necessary, return the slide to the iodine vapour until it is adequately stained. Avoid overstaining if the state of starch is of interest as this will mask birefringence of any granules in the early stages of gelatinization. Staining usually takes from 30 sec to up to 3 min depending on the nature of the specimen and the staining conditions (fuller staining is appropriate for protein).

5. When adequately stained add 1 drop of either Euparal or mineral oil and a coverglass. Examine in brightfield and between crossed polars. The results are similar to those given by aqueous iodine staining.

6.3.3 The use of trypan blue

Iodine solution is generally chosen for demonstrating starch but, unless a critically small amount is used, iodine stains both intact and gelatinized starch. This staining can obscure the birefringence of intact starch. Trypan blue has the merit of colouring damaged starch granules but leaving intact granules unstained so that their birefringence is undimmed by the masking effect of staining.

Trypan blue is a direct cotton dye with a structure similar to that of Congo red, first used by Jones (1940) to stain starch that had been damaged mechanically during milling. Both are cotton dyes which colour cellulose by a hydrogen-bonding mechanism which is facilitated by the elongated shape of the dye molecules. A similar mechanism probably explains the ability of these dyes to stain damaged starch. Both mill-damaged and moist heat-damaged (i.e. gelatinized) starch granules swell considerably in water, thus allowing the dye molecules to enter the granule and align with the starch polymers. Intact granules only absorb a small amount of water so remain unstained and can show their full birefringence.

Both trypan blue and Congo red stain protein, some of which may be closely associated with starch, but the trypan blue colouring of protein is less obtrusive than that of Congo red and this makes trypan blue the preferred stain for damaged and gelatinized starch present in flour and flour-based products.

Trypan blue as a stain for damaged starch. The dye content of commercial samples of trypan blue varies from 40 to 90%. Take sufficient to give 0.25 g of actual dye and dissolve in 100 ml of distilled water. If the dye content is unknown, prepare a 0.5% aqueous solution and subsequently dilute this if the staining appears too strong. The dye solution keeps well.

Specimen preparation. The trypan blue stain can be used on cryo-sections, smears, squashes or powders. It is particularly useful for following the progress of the cooking of starch-based sauces by staining successive smears of sauce as it thickens (see Chapter 8).

Staining method. Using a dropper, place sufficient stain on the pre-paration to cover it completely. Leave for 1–3 min, apply a coverglass, invert and blot on fluff-free blotting paper. Examine by brightfield illu-mination and between partly crossed polars. The results are summarized in *Table 6.3*.

Table 6.3:   Trypan blue-stained starches, protein and cellulose

Damaged starch granules	Blue (no birefringence)
Starch which has lost granular structure, e.g. that present in wafers and extruded products	Pale blue
Intact starch	Unstained, birefringent
Protein	Blue
Lignified cellulose	Dark blue
Cellulose	Pale blue

Trypan blue as a stain mountant for moulds. The main structural constituent of fungi is chitin which is similar in structure to cellulose, and so has an affinity for trypan blue. A simple stain mountant which is useful for showing the presence of moulds in sauces and tomato purées can be made from the trypan blue solution used to demonstrate damaged starch.

The stain mountant is prepared by dissolving 1 g of phenol in 30 ml of distilled water and adding 30 ml of glycerol (37.8 g) and then 40 ml of 0.25% aqueous trypan blue solution. The stain mountant keeps well.

For specimen preparation, make a thin smear of the sauce or purée. Add 2–3 drops of stain mountant to the specimen. Leave for 5 min. Apply a coverslip, invert and blot to remove excess mountant. Examine with 20× and 40× objectives.

Mould hyphae will stain blue and cellulose structures (plant cell walls) stain very pale blue, whilst protein stains a darker blue.

References

Flint FO. (1982) Light microscopy preparation techniques for starch and lipid containing snack foods. *Food Microstruct.* **1**, 145–150.

Flint FO, Firth BM. (1988) A Toluidine blue stain mountant for the microscopy of comminuted meat products. *Analyst* **106**, 1242–1243.

Flint FO, Pickering K. (1984) Demonstration of collagen in meat products by an improved picro-Sirius Red polarisation method. *Analyst* **109**, 1505–1506.

Hotchkiss RD. (1948) A microchemical reaction resulting in the staining of polysaccharide structures in fixed tissue preparations. *Arch. Biochem.* **16**, 131–141.

Jensen WA. (1962) Pectin hydroxylamine–ferric chloride reaction. In *Botanical Histochemistry*. W.H. Freeman, London, p. 202.

Jones CR. (1940) The production of mechanically damaged starch in milling as a governing factor in the diastatic activity of flour. *Cereal Chem.* **17**, 133–169.

7 Fat in Food

7.1 Introduction

Fat plays an important part in the microstructure of many processed foods. Although neutral fats have no flavour of their own, their presence and the way they are dispersed affects the taste and mouth feel of the food which contains them. This is because the flow properties of the food in the mouth (spreadability, coating of the tongue, viscosity and the sensation on swallowing) are greatly influenced by the fat fraction. Many foods, for example milk, cheese, fat spreads, ice cream and salad cream, taste richer when they contain a high proportion of fat, but for health reasons there is now a demand for low-fat versions of these products. In developing such products, the demonstration of fat and the materials used to replace all or part of it (e.g. starches and hydrocolloids) is important.

7.2 Polarization microscopy for solid fats

Fats which are solid at room temperature can be seen by polarization microscopy which demonstrates the crystalline nature of solid glycerides. Proof that the birefringence is due to fat can be found by warming the slide just sufficiently to melt the fat and then examining it whilst still warm. The molten fat will have lost birefringence which returns as the slide cools; the new crystals often have a different form to the original, usually being larger and sometimes forming spherulites (see *Figure 13.1*).

7.3 Oil soluble dyes – uses and limitations for demonstrating fatty food constituents

The most widely used methods for liquid fats in food (including mineral oil) are the oil-soluble dyes Sudan IV (Scarlet R), Oil Red O and Sudan Black B. These dyes colour liquid fats by a partition mechanism, the dye being more soluble in unsaturated glycerides (i.e. liquid fats) than in the aqueous solvent (alcohol or isopropanol) used as a staining medium. The presence of water in the solvent reduces fat loss from the section being stained and, by hydrating proteins and other non-fat constituents, prevents these from staining (Horobin, 1988). Unless warmed, solid (i.e. saturated) glycerides are not stained but, because they are crystalline, they can be shown by polarization microscopy. In foodstuffs, solid fats usually have liquid fats associated with them, so that by viewing a stained preparation through partially crossed polars both solid and liquid fats can be seen.

The speed and intensity of fat staining is dependent on the concentration of oil-soluble dye in the aqueous solvent used. Sudan IV and Sudan Black B are used as saturated solutions in 70% aqueous ethanol and Oil Red O as a supersaturated solution in 60% aqueous isopropanol (propan-2-ol), a method introduced by Lillie and Ashburn in 1943. In this technique, supersaturated Oil Red O in aqueous isopropanol is prepared by adding water to a stock saturated solution of the dye in isopropanol and filtering after 10 min. This method, which often leads to dye precipitation during staining, was improved by Cataloni and Lillie (1975) who added 1% dextrin to the water used in preparing the supersaturated dye solution. The dextrin used by Cataloni and Lillie had the effect of eliminating dye precipitation and also of substantially increasing the concentration of Oil Red O in solution. The type of dextrin used was not specified and, since the term 'dextrin' now covers a wide range of products made from different starches and hydrolysed to different extents, it is not surprising that commercial dextrins vary in their ability to prevent dye precipitation, and to give dye-rich solutions. The most suitable dextrin seems to be a moderately hydrolysed corn starch dextrin. This is one which gives a blue–purple colour with iodine solution; dextrins which are hydrolysed to the extent that they give little or no colour with iodine are quite unsuitable.

7.3.1 Staining with supersaturated Oil Red O
(adapted from Cataloni and Lillie, 1975)

Specimen preparation: 10- or 12-µm cryosections. For smears and squashes, see the rapid staining method (Section 7.3.2).

Reagents: The Oil Red O stock solution is prepared as follows:
- Oil Red O (CI 26125) 0.5 g (for best results use a dye certified by the Biological Stain Commission)
- Propan-2-ol 100 ml.

Gently reflux the Oil Red O in the propan-2-ol for 60 min and filter whilst hot using a coarse filter paper (e.g. Whatman no. 4). The stock solution keeps for several weeks.

The Oil Red O working solution is made by mixing together 60 ml of Oil Red O stock solution and 40 ml of 1% freshly prepared aqueous corn dextrin. Cover and allow to stand for at least 15 min, filter just before use. Change the solution daily.

Standard staining method for cryosections.
1. 60% propan-2-ol 1 min.
2. Oil Red O working solution 10–15 min.
3. Differentiate in 60% propan-2-ol for 1 min.
4. Rinse very gently in a beaker containing 400 ml water.
5. Wipe the underside of the slide, blot round the stained section and mount whilst still wet.

Mounting the stained section. This calls for great care if displacement of loose fat is to be avoided. Aquamount, Farrant's medium and aqueous glycerol (30% w/w) are all suitable mountants, as is the toluidine blue stain mountant described earlier (Section 6.3.7).

Glycerol–gelatin, the standard mountant for lipid-stained preparations, can cause movement and melting of the loosely held and oily fats present in some foods. It should only be used if the fat present is in the form of adipose tissue or is held within plant cells.

Liquid fat stains red by this method whilst solid fat is unstained but visible as birefringent crystals closely associated with the stained fat and seen when the slide is viewed between partially crossed polars.

7.3.2 A rapid staining method using Oil Red O

This rapid staining method may be used on cryosections, smears or squashes. It is particularly suitable for smears which can lose material in a multistage staining method.

Method. Prepare the Oil Red O working solution by mixing 15 ml stock of Oil Red O with 10 ml of 1% aqueous corn dextrin. Cover and allow to stand for 10 min. Meanwhile position one or two slides containing cryosections or smears in the base of a large Petri dish. Filter the working solution of Oil Red O through a Whatman no. 1 filter paper, reject the first few millilitres but, when the filtering starts to slow down, allow 2 – 3 drops of filtrate to cover the specimen. Replace the Petri dish lid immediately and allow the sections to stain for 5–10 min.

The next stage can involve rinsing and mounting as described in the standard staining method or, alternatively, for speed, the section can be mounted in the stain solution after draining away the excess. Mounting in the stain solution is recommended for very loosely held fat which may be lost in the water rinse. It also has the advantage that unstained and colourless structures are particularly easy to see because the stain solution has a lower refractive index than any of the mountants recommended.

7.3.3 A rapid staining method using Sudan IV and Oil Red O

An alternative to the Oil Red O supersaturated isopropanol method relies on the finding by Kay and Whitehead (1941) that mixed red Sudan dyes give saturated aqueous alcoholic solutions containing more colour than single Sudan colours. A stock staining solution made by saturating 70% ethanol with both Oil Red O and Sudan IV (Scarlet R) keeps well and can be filtered straight on to sections or smears contained in Petri dishes, as described in the rapid Oil Red O isopropanol method. The staining time and mounting of the stained sections can be exactly the same as that described in the rapid Oil Red O isopropanol method or, alternatively, the stained sections may be mounted in the toluidine blue mountant after lightly blotting away the excess staining solution.

7.4   Staining fats with osmium tetroxide vapour

With some foods, for example emulsions and products containing loosely held oily fats, solvents have to be avoided as the fat may be lost in the dye bath or displaced during staining or mounting. Staining with osmium tetroxide vapour overcomes these problems. Osmium tetroxide is not a dye but an unstable oxide which is reduced to brownish coloured lower oxides by the ethylene bonds present in unsaturated fatty acids.

Osmium vapour staining locates the distribution of liquid fats present in foods precisely and so provides a check on the results obtained using oil-soluble colours. However, because of the hazards associated with osmium tetroxide (see safety note below) and also its relatively high cost, the oil-soluble colours should be used in preference to osmium tetroxide vapour once it has been established that these give satisfactory results. This leaves food emulsions as the one type of food which usually, but not always, requires vapour staining.

Safety note: osmium tetroxide is a hazardous material. Its vapour is a fixative that can affect living tissue, including the cornea of the eyes. It is imperative that only small quantities of osmium are used and that these are handled with great care. Safety spectacles and gloves should be worn and staining *must* be carried out in a fume cupboard.

Preparation of osmium tetroxide vapour staining chamber. Set up a vapour staining chamber as described for iodine vapour staining (see Section 6.3.2) but using dry filter paper and use a fume cupboard for all operations. Pour a 5-ml ampoule of 2% osmium tetroxide (obtainable from Taab Laboratories) over the filter paper on the base of the jar and replace the lid immediately. Leave the emptied ampoule at the back of the fume cupboard and later dispose of it carefully. Leave for 5 min to allow the vapour to reach equilibrium. The chamber is now ready for use. It must be kept in the fume cupboard at all times.

Staining procedure. Smears or cryosections may be used. Place one or two prepared slides in the chamber with the specimen facing the osmium tetroxide-treated filter paper and quickly replace the lid. Staining takes from 1 to 5 min, and can be seen through the glass of the container as a browning of the specimen. The progress of staining can be checked by examining the unmounted slide using brightfield illumination and the 10× objective and, if crystalline fat is present, also viewing between partly crossed polars. If solid fat is present, avoid overstaining because, as staining proceeds, the light brown unsaturated fat becomes darker in colour and eventually black, which can obscure the birefringence of the fat. Several slides can be stained using one ampoule of osmium tetroxide solution but the chamber's activity steadily decreases, staining takes longer and, after about 1 h, requires renewal by the addition of a further ampoule of osmium tetroxide.

Mounting the stained preparation. Just before mounting, examine the section or smear so that any adverse changes that the mountant or the coverslip subsequently make to the specimen will be known. Then mount the specimen in a minimum quantity of 3-in-1 oil or 30% aqueous glycerol by placing one drop of mountant on a coverglass and gently lowering the specimen on to the mountant. When the specimen touches the mountant, invert the slide and allow the coverslip to settle. The slides can be stored for several weeks if kept covered in a refrigerator.

Examine the slides in brightfield and between partly crossed polars.

Unsaturated fats will stain light to dark brown or black and saturated fats are unstained. Birefringent crystals associated with stained fat are best seen in lightly stained preparations.

7.5 Combined osmium and iodine vapour staining

The vapour staining of both fat and starch can be done sequentially provided that the lipid staining is carried out first (see Section 6.3.2 for iodine vapour staining).

References

Cataloni RA, Lillie, RD. (1975) Elimination of precipitates in Oil Red O fat stain by added dextrin. *Stain Technol.* **50**, 297–299.

Horobin RW. (1988) *Understanding Histochemistry.* Ellis Horwood, Chichester.

Kay WW, Whitehead R. (1941) The role of impurities and mixtures of isomers in the staining of fat by commercial Sudans. *J. Pathol. Bacteriol.*, **53**, 279–284.

Lillie RD, Ashburn LL. (1943) Supersaturated solutions of fat stains in diluted isopropanol for demonstration of acute fatty degeneration not shown by Herxheimer's technic. *Arch. Pathol.* **36**, 432–435.

8 Food starches

8.1 Physical structure of commercial starches

Starch occurs as insoluble granules in many plant tissues, and is widely distributed in nature; cereal grains, peas and beans and roots and tubers being particularly rich sources. The important commercial starches are those extracted from cereals (corn, rice, wheat) and from roots and tubers, that is potato, and cassava (tapioca) starch. Like all starches, the size and shape of commercial starch granules is characteristic of the plant source and this enables them to be identified microscopically (see *Table 8.1*).

Other physical features which assist in starch identification are the appearance of the starch when viewed between crossed polars, the position of the hilum and the presence of striations in the granules (*Figure 8.1*). The striations are due to the layers that make up the starch granule which are laid down around an internal point called the hilum. The hilum may be central or eccentric. Striations are not obvious in all starches but the position of the hilum can usually be seen in non-aqueous mounts of starch occurring as a point or sometimes as a star-shaped crack. This is useful for distinguishing rye from wheat starch, which has granules very similar in size and shape. Rye granules have a hilum split into 3–5 rays whereas the wheat hilum is only a central point (Wallis, 1965).

All intact raw starches are birefringent and show a characteristic Maltese cross when viewed between crossed polars. The hilum lies at the centre of the cross so that using crossed polars provides an easy way of finding its position in both aqueous and non-aqueous mounts (see *Figure 8.1*).

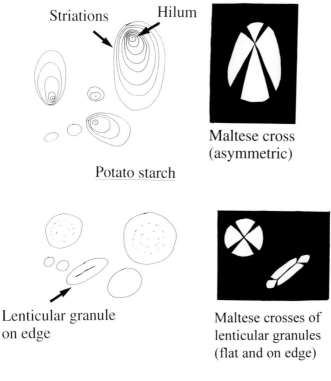

Potato starch

Wheat starch

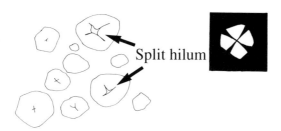

Maize (corn) starch

Figure 8.1: Diagrams of food starches as they appear when viewed with brightfield illumination and between crossed polars (approximately to scale).

When raw starch is damaged mechanically, as happens to a fraction of wheat flour during milling, the birefringence of the damaged granules is affected. Seen in an oil mount there is a loss of brightness with the arms of the Maltese cross appearing more diffuse but, when mounted in water, the damaged granules swell and lose much of this residual birefringence.

Rice starch

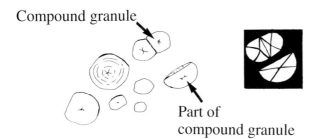

Compound granule

Part of
compound granule

Tapioca starch

Figure 8.1:   Continued.

Most samples of commercial starches contain a small amount of damaged starch, some of which may be due to heat damage caused during the drying of starch after separation by wet processing.

8.2   Chemical properties of commercial starches

Two main types of glucose polymer occur within starch granules. These are amylose, an essentially linear molecule with only limited branching, and the much larger, branching bush-shaped amylopectin molecule. Most starches contain about 25% amylose but tapioca starch contains only 17% and both maize and rice can occur in what is called a 'waxy' form which consists almost entirely of amylopectin. The term 'waxy' as distinct from 'floury' describes the appearance of the cut surface of the cereal grains (see *Figure 8.2*).

Table 8.1: Characteristics of commercial starches

Starch	Shape and size of main fraction	Appearance with crossed polars	Hilum	Striations
Maize (corn)	Polyhedral and rounded, 10–15 μm	Well-marked cross	Central star-shaped	None
Rice	Polyhedral and rounded, 5–8 μm	Well-marked cross	Central point	None
Wheat	Two size ranges: spherical, 3–7 μm	Weak cross	Central point	None
	lens-shaped 15–30 μm appearing circular when flat, biconvex when on edge	Weak cross when flat. Bright on edge with distinct line between arms of cross	Appears as a line when granules are on edge	Faint concentric on larger granules
Potato	Two size ranges: rounded 10–35 μm	Bright well-marked cross	Eccentric point	Faint concentric
	ovoid 30–100 μm	Very bright well-marked cross	Eccentric point near narrow end of granule	Well-marked concentric
Tapioca	Mostly rounded 10–30 μm characteristic flattened face on granules derived from compound granules	Well-marked cross	Central, split or star-shaped	Very faint concentric on larger granules

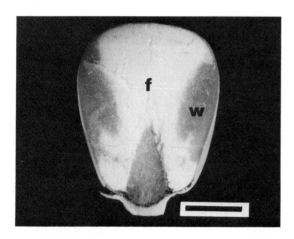

Figure 8.2: Interior of maize grain, f = floury, w = waxy zone. Bar = 2mm.

Waxy and floury starches behave differently on cooking, and extensive breeding of maize has produced starches with varying proportions of amylose and amylopectin. These include a starch containing 90% amylose.

Different amounts of amylose and amylopectin in a starch affect its reaction with iodine. In the presence of water, the long helical structure of amylose binds iodine to give a deep blue colour. This blue colour is given by normal starches even though they contain only about 25% amylose. The branches of the amylopectin molecule form only short helices so that waxy starches that have no amylose give only a dull reddish colour with iodine. This is similar to the colour given by hydrolysed starches (dextrins) which contain shortened amylose chains. This makes iodine a useful reagent for the identification of high amylopectin (waxy) starches and for following the changes that different starches undergo during cooking.

8.3   The effects of moist heat on starch

When heated in water to a sufficiently high temperature both normal maize and high amylose maize starch form a viscous paste which sets to a gel on cooling but when stored the gel shrinks and loses water. This loss of water is called syneresis. Syneresis is very marked if the gel is stored for any length of time. Such separation of water during storage can affect commercial lemon curd made from unmodified corn starch during the 6 months of its expected shelf-life. Syneresis is also accelerated by freezing. This results in starch-based sauces which have been frozen appearing curdled when they are cooked. This affects products such as frozen 'boil in the bag' fish in white sauce.

In contrast, amylopectin starches cook to give a viscous paste which flows but does not gel or suffer syneresis. This should make waxy maize starch ideal for preparing sauces that are to be frozen but the cooked paste made from unmodified waxy maize starch is very fragile. It loses viscosity if it is overheated or sheared and it also has a poor texture, being stringy and cohesive (mucus-like).

8.4   Starch gelatinization

The changes that starch undergoes during cooking can be followed microscopically. They involve a process called starch gelatinization which occurs to a greater or lesser extent whenever starch is cooked in the presence of water.

At room temperature, only damaged starch swells in water but, when the water is heated, a temperature is reached when the largest intact granules suddenly swell and lose birefringence. At higher temperatures progressively smaller granules swell and lose birefringence. The temperature range over which these changes occur is known as the gelatinization temperature which is characteristic for the type of starch involved (see *Table 8.2*).

Table 8.2: Gelatinization temperatures of food starches

Source	°C
Potato	58–68
Wheat	57–64
Tapioca	55–67
Rice	67–78
Maize	62–72

The temperature ranges shown in *Table 8.2* vary with the history of the starch samples used, and the conditions of the test. Slightly different ranges can be found in the literature. More significant differences in gelatinization temperatures are shown by starches with high amylose contents. Whereas normal and waxy maize starch gelatinizes in the range 62–72°C or slightly above, a high amylose maize does not even swell until a temperature of about 85°C is reached, and it takes several minutes of boiling for birefringence to be completely lost.

8.4.1 The gelatinization process

The gelatinization process can be followed microscopically and the gelatinization temperature determined by heating a suspension of starch in water. Potato starch is recommended for a first experiment. The granules are large, making them easy to see with the microscope and their initial birefringence is bright, so that during the heating process the reduction in birefringence as the granules swell is easily followed. Granules can often be seen which are partly swollen and partly intact. Trypan blue is recommended as a stain for the granules because it colours only the swollen granules, leaving the intact ones unstained so that their birefringence can be clearly seen. Iodine, in the form of iodine vapour, can be used to show the loss of amylose by the granules (see *Figure 8.3f*).

Materials needed for the experiment.
- Sample of commercial grade potato starch
- Distilled water
- Four boiling tubes
- Four dipping tubes (internal diameter 2 mm)
- Water bath

- Thermometer
- 0.5% aqueous trypan blue

Method.
1. Weigh 400 mg of starch into each tube and add to each tube 10 ml of water.
2. Shake one of the tubes and use a dipping tube to deposit 1 drop of the suspension on a microscope slide.
3. Add 1 drop of aqueous trypan blue and mix.
4. Apply a coverglass, blot away any excess solution and examine with a 10× objective in brightfield and between crossed polars. This will show the full birefringence of most of the granules with just the occasional stained granule showing the presence of damaged starch.
5. Meanwhile, heat all four tubes to about 45°C, shake the tubes and adjust the heating to no more than 2°C/min.
6. Continue heating and occasionally shaking the tubes to maintain the starch in suspension.
7. Remove individual tubes at 58, 63 and 68°C, cooling each tube as it is removed.
8. Leave the last tube to reach a temperature above 90°C and stir this one vigorously.
9. Sample and stain each of the suspensions as described and examine in brightfield and with crossed polars.

Results *(Figure 8.3)*. From the 58°C tube *(Figure 8.3a,b)* note the 'bold' outline shown by unstained granules, indicating a refractive index difference between the starch and water. Swollen granules are coloured and have a 'softer' outline. The 68°C tube *(Figure 8.3e,f)* shows gelatinized granules which are swollen but have a collapsed 'imploded' appearance. Crossed polars show the progressive loss of birefringence throughout the starch gelatinization range.

Note the resistance of the smallest granules to gelatinization, some of which are still birefringent at 68°C. The gelatinization temperature for starches is usually taken to be the temperature range needed for the gelatinization of 90% of the granules. The sample heated to above 90°C shows a reduction in viscosity after shearing and many of the sheared granules are ruptured, as can be seen in the trypan blue preparation.

During gelatinization, amylose is released by the starch granules. This is important because this free amylose enables the cooked starch to gel. If facilities are available for iodine vapour staining, stain a smear from the 68°C tube to show this free amylose (see *Figure 8.3f*).

When all the tubes have been examined, stopper them and leave overnight in a refrigerator. After storage, the 68°C and above 90°C tubes show the effects of syneresis: the gel which forms shrinks and a watery exudate is produced.

The observation of starch gelatinization by microscopy goes some way towards explaining the cooking behaviour of starches containing different

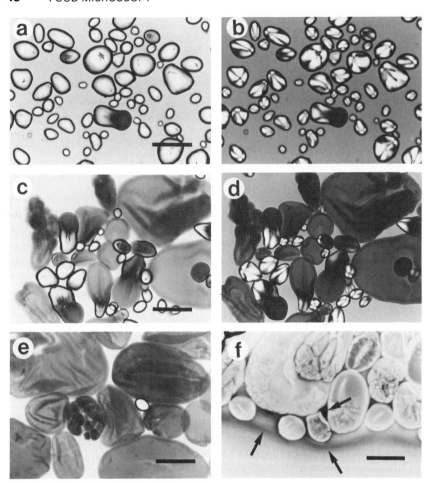

Figure 8.3:   Stages in the gelatinization of potato starch. (a)–(e) Smears of heated starch suspension stained with trypan blue, viewed brightfield and with partly crossed polars. (a) and (b) At 58°C starch is beginning to swell, the swollen granules take up stain as they lose birefringence (b). (c) and (d) At 63°C half the granules are swollen, imploded, and stained, (d) shows birefringence lost by stained granules. (e) At 68°C the starch has reached the gelatinization end-point. Note the remaining intact granule (central). (f) A smear of 68°C starch suspension stained with iodine vapour shows amylose exudate (arrows). Bar = 100 μm.

amounts of amylose. The gels formed by amylose-containing starch are due to the release of amylose. Amylopectin starches yield a viscous paste as the gelatinized granules absorb water and remain suspended. This paste does not gel, making it suitable for sauces, but the suspended swollen granules are fragile. This explains their need for chemical modification.

8.5   The microscopy of chemically modified starches

Waxy maize starch is strengthened by the formation of chemical links at random positions within the raw granules. This 'spot welding', as it has been called, has a dramatic effect on the cooked granules, making them resistant to shear so that the thickened paste can be made into sauces that withstand all the mechanical operations involved in manufacturing frozen meals and canned products.

In the raw state, that is when ungelatinized, cross-linked starches react poorly with iodine, giving only a straw or light brown colour with dilute iodine. When cooked, modified starch gives a reddish purple with iodine. Good examples in which to see this cooked starch are commercial salad creams and the tomato sauce in which baked beans are canned. A comparison of the starch present in the cooked beans with that in the tomato sauce shows how starch gelatinization is affected by the availability of water. The starch in the cotyledon cells which form the bulk of the bean still shows some birefringence, whereas the modified corn starch in the sauce, which has access to water throughout processing, is fully gelatinized, and may even show some disruption (see *Figure 8.4*).

8.6   Other modified starches

The term 'modified starch', widely used in food labelling, covers a range of treatments and includes pre-gelatinization and acid-thinning of starches. Pre-gelatinized starches are used in custard powders and for instant desserts (see *Figure 8.5*). Acid-thinned starches are used in sugar confectionery to make starch jellies, such as Turkish Delight.

The traditional type of Turkish Delight which is sold with a starch and sugar dusting consists of a flavoured, sweetened starch jelly. It is made from thin boiling corn starch, that is starch which has been treated with 0.5% hydrochloric or sulphuric acid. This makes the starch suspension much more fluid when it is boiled with sugar and liquid glucose. It still requires a lot of cooking because the effect of the sugary ingredients is to raise the gelatinization temperature of the starch. The solids content before the jelly can be poured is 75–80%, most of which is sugar which, during cooking, competes with the starch for the available water.

On storage, traditional Turkish Delight suffers syneresis, that is the jelly expels water. This is why it is packed in corn starch and icing sugar which absorb the exuded moisture to form an outer crust.

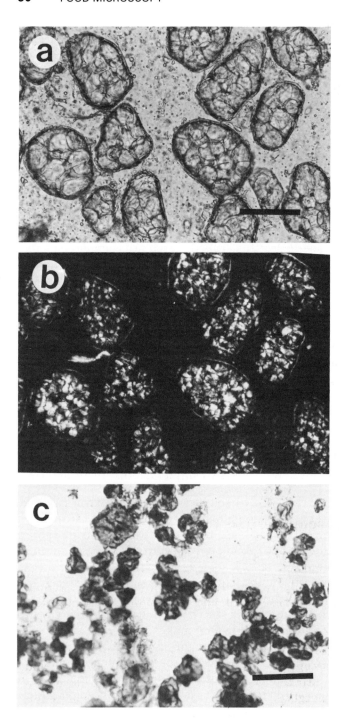

Figure 8.4:   Canned beans in tomato sauce. (a) Bean cotyledon cells viewed in brightfield. (b) Same field as (a) viewed between crossed polars. (c) Modified corn starch in sauce stained with aqueous iodine. Bar = 100 µm.

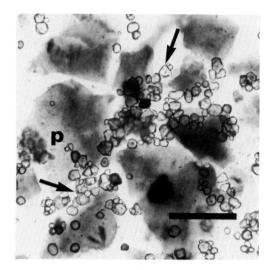

Figure 8.5: 'Instant' custard powder stained with aqueous iodine. p = pre-gelatinized starch, arrows show chemically modified starch. Bar = 100 µm.

Traditional Turkish Delight, made as described above, cannot be coated with chocolate because the syneresis water would dissolve the sugar in the chocolate (chocolate contains about 50% ground sugar). Chocolate-coated Turkish Delight usually consists of a pectin jelly which contains some cooked starch. The starch gives the jelly the translucence characteristic of Turkish Delight, whilst the pectin acts as a gelling agent which can absorb any water released by the cooked starch.

Recently, chocolate-coated jellies made from chemically modified starch have become available. A comparison of the two types of chocolate coated jellies with the starch-dusted Turkish Delight makes an interesting experiment.

8.6.1 The microscopy of Turkish jellies

The aims of the experiment are to discover the extent of starch gelatinization in each jelly and to identify the pectin-containing jelly. The reagents needed are distilled water, dilute iodine (5 ml of 1% iodine in 2% potassium iodide diluted to 100 ml) and the toluidine stain mountant (see Section 6.3.1).

Starch demonstration. Place a piece of jelly 2–3 mm in diameter on the centre of a microscope slide. Add one drop of distilled water and, using a needle, gently stir the jelly until it is dispersed. Add one drop of dilute iodine and a coverglass. Examine with the 10× and 20× objectives using brightfield illumination and crossed polars. Repeat for each sample.

Note the relative amounts of starch present and the extent of starch gelatinization. Observe the 'imploded' and 'shrivelled' appearance of some of the cooked starch (see *Figure 8.6*).

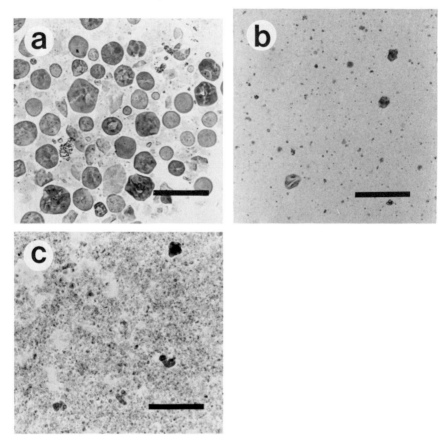

Figure 8.6:   Turkish Delight jellies stained with aqueous iodine. Bar = 100 μm.
(a) Chocolate-coated jelly based on modified starch; starch is gelatinized but still granular.
(b) Chocolate-coated jelly based on pectin and modified starch; starch is gelatinized and
almost completely disrupted. (c) Starch-dusted jelly based on modified starch; starch is
almost completely disrupted, the heavier staining is due to a higher starch content than
that present in (b).

Some raw (i.e. birefringent) starch may be seen in any of the samples.
This is derived from the coating of the authentic Turkish Delight or the
moulding starch used to make the jelly pieces for chocolate coating.

Pectin demonstration. Disperse a small amount of jelly in one drop of
water and add one drop of the toluidine stain mountant. Leave for 1 min,
add a coverglass, invert and press gently on blotting paper to remove
excess stain. Examine with the 10× and 20× objectives.

Note the magenta colour produced by the pectin-containing jelly and
compare this with the pale blue colour given by the starch jellies. The
magenta colour is produced by '*in situ*' dye polymerization which is only
possible when the dye-binding sites are close together.

8.7 Starch in baked products

The effects of added ingredients on the gelatinization of starch are important because they help to explain why the small range of ingredients (flour, fat, sugar) used to make baked products produces such a variety of textures. Much of the textural difference shown by cakes, pastry and biscuits is due to the state of the starch granules they contain (see Table 8.3).

Table 8.3: State of wheat starch in baked foods

Product	State of starch granules			
	Intact	Swollen	Gelatinized	Disrupted
Bread		←——————→		
Cake (e.g. Madeira)		←——————→		
Pastry (short crust) outer layers	←——→			
interior		←——————→		
Cream cracker outer layers	←——→			
interior		←——————→		
Short sweet biscuit (e.g. shortcake) outside	←——→			
inside		←——→		
Ice cream wafers containing sugar			←——————→	
containing saccharin			←——————→	

The reason for the different extents of starch gelatinization shown in Table 8.3 is the availability of water. Sugar restricts water availability and so does fat (by coating the starch granules). This explains why the starch in cakes is less gelatinized than that in bread, even though a cake batter contains more water than does a bread dough.

The reason for raw starch in the outer layers of pastry and biscuits is that the uncooked products contain only a limited amount of water, this allows the outer layers to dry out in the oven before they can reach gelatinization temperature.

Protein is important in bread structure where the gluten provides the carbon dioxide-retaining structure in the early stages of baking. Later, the heat that denatures the protein gelatinizes the starch which then helps to form the final structure. In cakes, egg protein is initially structure forming but the stability of the final cake depends largely on the presence of uniformly swollen starch granules.

The ice cream wafer has a totally different structure. Instead of a protein matrix holding gelatinized starch granules, a matrix of gelatinized starch which has largely lost its granularity entraps air, bran, oil droplets and protein particles. This sort of structure is also typical of many extruder-cooked products including crispbreads, breakfast cereals and snack foods.

8.7.1   The microstructure of baked products

Practical aspects. The best way to study the state of starch in food products is to work with thin sections, that is 10–12 µm cryosections or 8–10 µm wax sections. (Wax sections can be cut in a cryostat with the temperature set at 0°C.) The sections may be stained with iodine vapour or mounted in toluidine blue stain mountant to show the unstained starch in relation to other constituents (see *Figure 8.7*).

However, the lack of sectioning equipment should not deter the food microscopist from studying the state of starch in what is a wide range of readily available important food products. The microstructure of baked goods makes an interesting topic, because the extent of wheat starch gelatinization varies both between different products and also within the products themselves, as shown in *Table 8.3*.

8.7.2   Preparation of baked foods for the microscope

Bread and cake. Soak in water a crumb taken from the centre of the loaf or cake. Take the wet crumb in forceps and brush it about on the centre of the microscope slide. This will detach starch granules from the crumb without damaging them. Add one or two drops of either dilute iodine or 0.5% aqueous trypan blue, leave for 1 min before applying a coverglass. View with the 10× objective.

Bread shows imploded, swollen starch granules some of which appear wrinkled. Cake shows swollen starch granules which are less swollen than those in bread and which have a bolder outline showing them to be less gelatinized (*Figure 8.8*).

Baked foods with a low moisture content. Many baked foods, including short sweet biscuits and crackers, are crisp, indicating a low moisture content. These benefit from overnight storage in a moist atmosphere before sampling (e.g. in a Petri dish containing moist filter paper with the specimen placed on a slide so that it does not come in contact with the wet paper). This serves to swell the product, which makes sampling of the outer and inner layers easier; for example the layers which make up cream crackers can now be peeled apart. Take slivers of the outer and inner layers of the biscuit and transfer these to an embryo dish or watch glass containing water. Next, impale a sliver on a needle and brush it about on a microscope slide to detach starch granules. Stain with trypan blue as described above (avoid iodine staining as this can easily mask residual birefringence). View starch from the different layers using partly and fully crossed polars. Note the more gelatinized state of starch from the interior of the product.

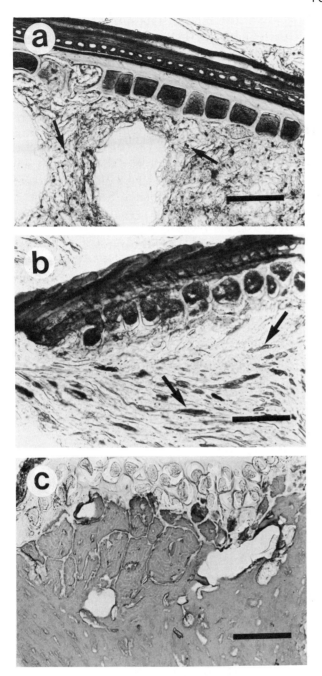

Figure 8.7:   Cryosections (12 μm) of baked and extruder-cooked products. Bar = 100 μm. (a) Wholemeal bread stained with toluidine blue mountant showing bran fragment and yeast cells (arrows). (b) Extruder-cooked wholemeal crispbread stained with toluidine blue mountant showing bran fragment and gluten particles (arrows). (c) Extruder-cooked wholemeal crispbread stained with iodine vapour showing aleurone cells of bran entrapped in an almost amorphous starchy matrix.

Ice cream wafers. These are of two distinct types: those that contain an artificial sweetener (usually saccharin) and those made with sugar as an ingredient. The saccharin wafers are thin, single-layered, and rectangular in shape. The fan-shaped sugar wafers appear thicker but are made by folding four thin layers of biscuit together. Both types of wafer contain wheat starch in a very gelatinized state.

A comparison of the two types of wafer makes an interesting experiment which illustrates the effect sugar has on gelatinization. For this, moisten portions of the two types of wafer as described in the previous section. Take small slivers of each wafer, and transfer to slides containing a drop of water. Use two needles to pull the slivers into small pieces and transfer one or two of these pieces to another slide. Stain with one or two drops of either dilute iodine or 0.5% aqueous trypan blue. Leave for 1 min before applying a coverglass. View with the 10× objective. Note the presence of disrupted (amorphous) starch in both samples. The sugar wafer also contains starch which, though gelatinized, is still granular, whereas almost all the starch in the saccharin wafer is in an amorphous state (see *Figure 8.9*). Cryosections stained with iodine vapour are even better for showing the difference between wafers because additional swelling of granules can be largely avoided.

Extruded starch-based products. The method described for ice cream wafers can be adapted for extruder-cooked starch products such as breakfast cereals and crispbreads, but these too benefit from cryostat preparation.

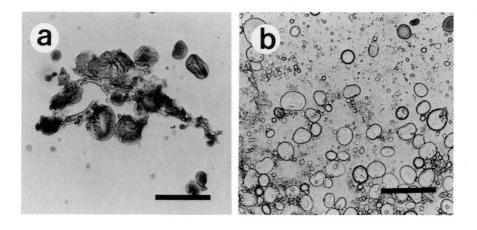

Figure 8.8: Starch granules from bread and cake treated with aqueous iodine. Bar = 100 µm. (a) The imploded, swollen starch granules from bread stain readily with iodine. (b) The starch from cake is less swollen, iodine staining has been largely inhibited by the fat present in the cake.

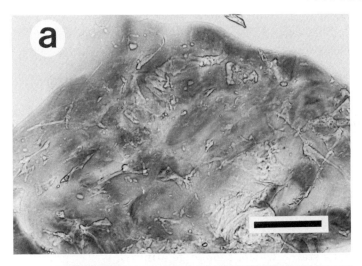

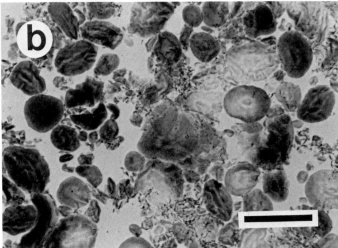

Figure 8.9: Wholemounts of ice cream wafers stained with aqueous iodine. Bar = 100 μm. (a) The sugar-free wafer consists largely of disrupted (amorphous) starch. (b) Most of the starch in the sugar-containing wafer is very swollen and imploded but is still in a granular state.

Further reading

Angold RE. (1979) Cereals and bakery products. In *Food Microscopy* (ed. JG Vaughan). Academic Press, London, pp. 75–136.

Evers AD. (1979) Cereal starches and proteins. In *Food Microscopy* (ed. JG Vaughan). Academic Press, London, pp. 139–186.

Flint FO. (1982) Light microscopy preparation techniques for starch and lipid containing snack foods. *Food Microstruct.* **1**, 145–150.

Jones CR. (1940) The production of mechanically damaged starch in milling as a governing factor in the diastatic activity of flours. *Cereal Chem.* **17**, 133–169.

Kent NL. (1983) *Technology of Cereals,* 3rd edn. Pergamon Press, Oxford.

Kent-Jones DW, Amos AJ. (1967) *Modern Cereal Chemistry*, 6th edn. Food Trade Press, London.

Lineback DR, Wongsrikasem E. (1980) Gelatinisation of starch in baked foods. *J. Food Sci.* **45**, 71–74.

Varriano-Marston E. (1982) Polarisation microscopy: applications in cereal science. In *New Frontiers in Food Microstructure* (ed. DB Bechtel). American Association of Cereal Chemists, St Paul, MN, pp. 71–108.

Wallis TE. (1965) Starch and bran. In *Analytical Microscopy* 3rd edn. J. and A. Churchill, London, pp. 67–82.

Watson SA. (1964) Determination of starch gelatinisation temperature. In *Methods in Carboydrate Chemistry,* Vol. IV (ed. RL Whistler). Academic Press, New York, pp. 240–242.

9 Meat, Fish and their Products

9.1 Meat

Meat can be regarded as the flesh of mammals and birds that is used as food. Most of the meat which is eaten consists of skeletal muscle with its associated connective tissue, including fatty adipose tissue.

9.2 Structure of skeletal muscle

Skeletal muscle contains bundles of minute fibres. Each nucleated fibre is held in its bundle by a fine network of connective tissue fibres called the endomysium. Each bundle of muscle fibres is surrounded by a layer of connective tissue (the perimysium) which is continuous with a thicker layer of connective tissue, the epimysium, which surrounds the muscle itself.

Muscle fibres are long, slender and multinucleated, the length varying from a few millimetres up to 34 cm. They may stretch from one end of a muscle to the other, although they are only about 10–100 μm in diameter (Bailey, 1972). The nuclei lie in the sarcoplasm immediately below the sarcolemma (cell membrane) which contains many long myofibrils each about 1 μm thick which make up the bulk of the muscle fibre. The electron microscope shows these to have a banded appearance due to the arrangement of actin- and myosin-containing filaments within each myofibril and this banding gives the fibre its striated appearance when viewed with the light microscope (*Figure 9.1*).

As well as skeletal muscle, there are two other types of muscle tissue which are eaten as food. These are cardiac muscle and smooth, unstriated or visceral muscle.

Cardiac muscle forms the bulk of heart muscle tissue and is classed as organ meat. Although they contain myofibrils similar to those in skeletal muscle, cardiac muscle cells show only faint striations and are characterized by their branching nature.

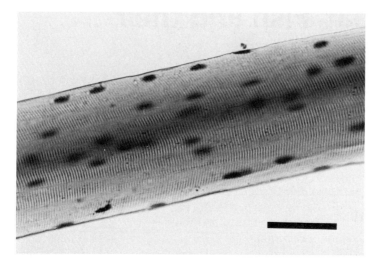

Figure 9.1: Wholemount of muscle fibre from bacon stained with aqueous toluidine blue. Note striations and dark stained nuclei. Bar = 50 μm.

Smooth muscle is quite different from skeletal and cardiac muscle. It consists of much smaller spindle-shaped cells thickened in the middle with narrow tapered ends. The central nucleus is elongated parallel to the long axis of the cell.

Smooth muscle occurs in blood vessels and the viscera, and it is an important constituent of tripe, which is prepared from the stomach lining of the cow.

9.3 The skeletal muscle fibres of meat

Skeletal muscle fibres vary considerably in size, which increases with age, use and hormone treatment. In meat from a very young animal, for example veal, the fibre diameter is only about 5 μm, while at maturity diameters are much greater, reaching 50 μm for mutton and 90 μm for pork (Lawrie, 1991). Although there is this size variation, it is not possible to identify the animal from which meat has come; poultry, beef and vertebrate fish all have similar skeletal muscle fibres which vary in size throughout the body. Fibre diameters also vary with processing, increasing with phosphate treatment and shrinking on cooking.

Striations are the most characteristic feature of skeletal muscle fibres. When meat has been canned, the nuclei are lost and much of its structure altered but striations persist even in damaged fibres although they are then less clear and often distorted, appearing like faint fingerprints.

Striations are most easily seen in wholemount preparations of raw or lightly processed muscle fibres which have been teased apart to separate the muscle fibres from the surrounding connective tissue.

9.4   Preparation of wholemounts of striated muscle fibres from meat

1. Place a small piece of raw or cooked meat on a Petri dish and flood with water.
2. Pin down the meat specimen with a needle and with the tip of a scalpel prise away a few fibre bundles.
3. Transfer one or two of these to a microscope slide containing a drop of water and use two needles to tease apart the individual muscle fibres. A low power stereomicroscope makes this task easier.
4. Once the fibres are well separated, drain off excess water and add 1–2 drops of 0.05% aqueous toluidine blue and leave for at least 5 min before putting on a coverglass.
5. View with 10× and 20× objectives using both brightfield illumination and crossed polars.

The fibres of raw meat stain a pale blue with the cell nuclei purple. The striations are more obvious if the aperture diaphragm is closed down a little and the polars are fully crossed. The nuclei may be missing from cooked meat although they survive in some smoked products, such as smoked dry cured bacon which makes an excellent subject for demonstrating muscle striations and nuclei (*Figure 9.1*).

9.5   Cryosections of meat

Preparing satisfactory cryosections of muscle tissue when it is in the form of meat can present problems. These are due to the physical state of the meat. Meat that has been previously frozen will have suffered damage caused by the relative slowness of commercial freezing. Slow freezing causes water to be withdrawn from the tissue, which becomes dehydrated as the water freezes. On thawing, some water returns to the muscle fibres but they never fully rehydrate because their original structure has been damaged. Unless the freezing process itself is being studied, it is recommended that frozen meat is fully thawed before preparing it for the cryostat.

Polyphosphates cause another problem. These are sometimes added to meat, especially chicken, bacon and ham, to increase its water-holding capacity, with the result that the muscle fibres are in a swollen state. Even without phosphate treatment, muscle has a high moisture content, containing about 75% water. This explains the need for very rapid freezing of fresh meat when preparing it for the cryostat.

Meat that has been cooked is much easier to section and is recommended as a material for learning the methods of meat sectioning.

Quenching in liquid nitrogen is recommended for meat but if other quenching methods are used then the sample size should be kept as small as possible, even if this means preparing several blocks for sectioning.

In setting the section thickness the orientation of the muscle fibres should be considered. If the fibre bundles are to be cut transversely then 10–12 µm sections are satisfactory. These can be stained to show cross-sections of the muscle fibres with their associated cell nuclei and surrounding connective tissue. (*Figure 9.2*). If the fibre bundles are to be cut longitudinally with the intention of seeing muscle striations (using only a simple staining technique), much thicker sections are needed, unless the meat is from a very young animal. This is because thicker longitudinal sections provide a greater proportion of the fibre thickness so that striations are more prominent and their birefringence is correspondingly brighter. With pork, having muscle fibres up to 90 µm in diameter, sections of 20 µm thickness are not inappropriate.

By cutting and staining sections of different thicknesses, an optimum thickness can be found for each of the constituents present in the sample of meat being studied. The constituents include the various connective tissues associated with meat, all of which may be seen when cryosections stained with the toluidine blue stain mountant are viewed in brightfield and between crossed polars.

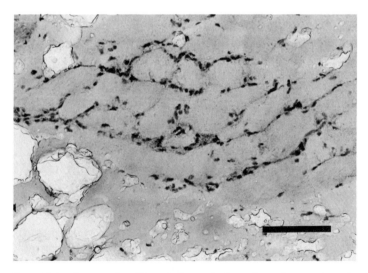

Figure 9.2:   Cryosection (12 µm) of muscle tissue in pork sausage stained with toluidine blue stain mountant. Most of the fibres are in transverse or oblique section. Bar = 100 µm.

9.6   Connective tissues associated with meat

In the Meat Products and Spreadable Fish Products Regulations 1984 (SI 1984 No 1566) meat is defined as 'the flesh including fat and the skin, rind, gristle and sinew in amounts naturally associated with the flesh used'. These associated connective tissues feature as ingredients in meat products, making their identification of some importance.

9.6.1   Fat

Fat occurs as adipose tissue, which is a specialized form of connective tissue in which membrane-bounded fat cells occur within a network of fine collagenous fibres. As a meat animal grows, fat is first deposited round the internal organs (mesenteric or flare fat), then under the skin where, in pigs, it forms head and back fat. Finally, fat deposits are laid down in the perimysial spaces within skeletal muscle where it is known as 'marbling'.

The essential structure of meat adipose tissue as seen with the light microscope is that of a three-dimensional network of collagen fibres filled with a mixture of solid and liquid fats. The size of fat cells varies according to location, being largest in the adipose tissue which is first laid down, that is in mesenteric fat where the cells are often greater than 100 μm in diameter. The hardness of the fat varies with location; mesenteric fat is highly crystalline whereas back and head fat is softer and less crystalline. The chemistry of the fat also varies with species; mutton fat is more saturated and therefore harder than beef fat, which in turn is harder than pork fat.

Identification of adipose tissue. Even the hardest and most crystalline meat fat contains some unsaturated liquid fat which can be coloured with lipid stains, and the crystalline fraction of the fat can be seen when the preparations are viewed between crossed polars. Both the fat and the collagen which surrounds the fat cells are seen easily in 12-μm cryosections of meat which have been mounted in the toluidine blue stain mountant (see *Figure 9.3*).

In comminuted meat products, the adipose tissue is present in small blocks often containing only a few cells. The collagen and fat in these can be seen both in cryosections and also in smears of raw products (e.g. sausage meat, see Section 9.8).

9.6.2   Skin and rind in meat products

Cattle hide is a source of gelatine and also of the collagen used to manufacture casings (sausage skins), but only pigskin, in the form of rind, is used as an actual meat product ingredient.

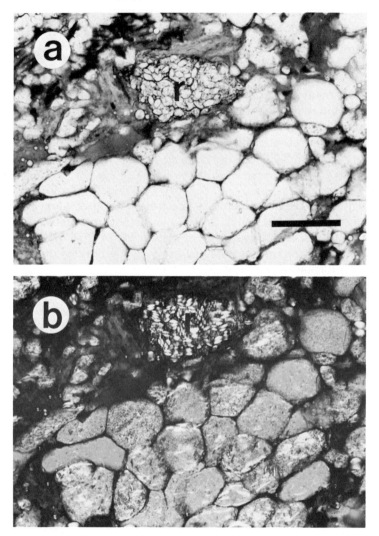

Figure 9.3: Cryosection (12 µm) of pork sausage stained with toluidine blue stain mountant. Bar = 100 µm. (a) Adipose tissue and rusk (r) seen with brightfield. (b) The same field as (a) viewed between partly crossed polars shows crystalline fat present in the adipose tissue and that the rusk starch is birefringent, i.e. raw.

Raw rind is pigskin which has been scalded to remove the epidermal tissue and bristles. It consists almost entirely of large bundles of parallel collagen fibres interwoven to produce a tough flexible three-dimensional structure. Running through this structure is a small amount of very fine, branched elastin fibres. Although raw rind is extremely tough, some of it is ground sufficiently fine to be used in comminuted meat products, but most rind is first cooked. It is heated in water at 80–100°C until soft enough for mincing and this minced rind forms an important ingredient of sausages and some canned luncheon meats.

During the cooking of the rind, the fibrous structure of the raw rind and the birefringence of its collagen fibres is progressively lost, but much of the rind found in raw meat products retains sufficient of its fibrous structure to make it easy to identify.

Identification of rind. The fibrous structure of rind can be seen in 10- or 12-µm cryosections of raw sausage meat stained with the toluidine blue stain mountant. The collagen fibres are coloured a pale lilac and the elastin fibres turquoise; the cells that produce collagen and elastin (fibroblasts) are coloured blue–violet. Collagen fibres are prominent and their birefringence greater in raw and lightly cooked rind. They appear more diffuse and show little or no birefringence after longer cooking but this makes the fine elastin fibres, which may be difficult to identify in raw rind, easier to see *(Figure 9.4)*.

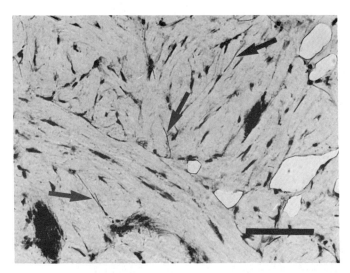

Figure 9.4:   Cryosection (12 µm) of cooked rind in pork sausage stained with toluidine blue stain mountant; very fine elastin fibres are just visible (arrows). Bar = 100 µm.

9.6.3 Tendons

The 'gristle and sinew' of the Meat Product Regulations includes tendons, which are tough fibrous structures that attach muscles to bones or other tissues. Tendon connective tissue is continuous with the connective tissue present in the muscle. Tendons can be seen most clearly if a whole muscle is excised from part of a small meat animal, such as a chicken or rabbit leg.

Identification of tendon connective tissue
1. Remove the skin from a fresh chicken leg. This will reveal entire muscles. Excise one of these including the tendon, that is the narrow part which joins

the bone. Transfer the muscle to a plate or large Petri dish and cut off about 1 cm of the tendon.

2. Moisten with water and use a scalpel and needle to separate the fibrous tissue into strands. Transfer pieces a few millimetres in length to a microscope slide and tease the strands into fibres using two needles.

3. Cover the teased fibres with the toluidine blue stain mountant and leave for 2 min. Drain off excess stain and position a coverglass over the stained fibres. Invert the slide and press on absorbent paper to further spread the stained fibres to give a thin preparation.

4. Examine with a 10× objective to view the bundles of birefringent pink-stained collagen fibres and the dark purple fibroblast cells (see *Figure 9.5*).

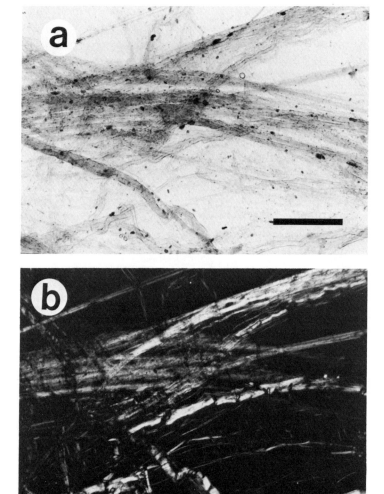

Figure 9.5:   Wholemount of teased tendon from chicken leg stained with toluidine blue stain mountant showing collagen fibre bundles. Bar = 100 μm. (a) Viewed with brightfield illumination. (b) Viewed between crossed polars.

9.6.4 Elastic tissue and ligamentum nuchae

Most connective tissue consists of collagen with only small amounts of associated elastin, but there are two sources of connective tissue which are particularly rich in elastin and both can be found in meat products. These are blood vessels, in which the elastic tissue fibres form concentric rings, and ligaments, where the fibres are arranged parallel to the axis of the ligament. In most ligaments the elastin is associated with collagen fibres, but in the bovine neck ligament (ligamentum nuchae) these are absent and the tissue consists of thick elastin fibres in a mucopolysaccharide matrix.

Unlike collagen which gelatinizes on cooking, elastin is extremely resistant to moist heat, and can be pressure cooked for long periods without suffering any change. Until comparatively recently, bovine ligamentum nuchae was a waste product but modern comminuters can now grind it sufficiently fine to enable it to be utilized in meat products. Dried ligamentum nuchae is also sold as a dog chew.

Identification of elastin fibres. Elastin fibres vary considerably in thickness. Finest of all are the slender fibres found in skin, being easiest to see in sections of well-cooked rind. Somewhat thicker fibres are found in arteries which are often opened up in meat processing so that the concentric waves of elastin fibres seen in the cross-sections of blood vessels present a layered structure. These are often seen in sections of canned meat products such as pork luncheon meat. The thickest elastin fibres are found in ligamentum nuchae.

All these forms of elastin stain readily with the toluidine blue stain mountant, giving a characteristic turquoise colour. The optimum cryosection thickness for seeing elastin fibres is 10–12 µm.

The identification of elastin in ligamentum nuchae. Fresh ox ligamentum nuchae may be obtained from some butchers, being known in the British meat trade as 'paddy wack'. This is easy material to prepare for the microscope. The fibres, which retain their elasticity, can be pulled apart very easily. Mounted in the toluidine blue stain mountant and allowed to stain for 2 min, the mucopolysaccharide matrix which surrounds the fibres stains pink due to the presence of chondroitin sulphate, and the fibres appears a greenish turquoise.

Dried ligamentum nuchae is sold in petshops as 'sinew'. The tissue has been dehydrated by oil cooking to give a hard fibrous amber-coloured product, sold as a dog chew. The dry product can be rehydrated by pressure cooking in an excess of water for 30 min. This yields a supple elastic material and an examination of the elastin fibres shows them to have been unaffected by oil cooking or the subsequent pressure cooking (see *Figure 9.6*).

Both the fresh and rehydrated dried ligamentum nuchae may be preserved by dehydration in two or three changes of acetone. The tissue dried in this way keeps well and hydrates readily.

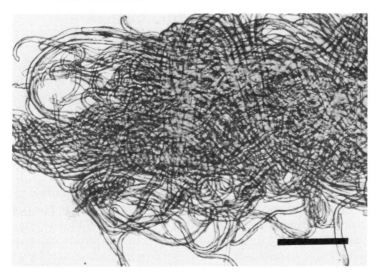

Figure 9.6: Elastin fibres teased from bovine ligamentum nuchae stained with toluidine blue stain mountant. Bar = 100 μm.

9.7 Starch constituents of meat products

Wheat rusk, that is sausage rusk, potato and corn starch, especially modified corn starch, are ingredients much used in modern meat products.

Sausage rusk is a chemically raised biscuit-like material baked from wheat flour. It contains both raw and gelatinized starch granules and is an important ingredient of British, as distinct from continental, sausages. The physical state of the starch is important.

The gelatinized starch present in the rusk readily absorbs water when the sausage is being prepared; on cooking the raw starch gelatinizes and then it is able to absorb the water lost by the meat ingredient when this is denatured by heating. This effect can be seen by comparing 10-μm cryosections of raw and cooked sausage (see *Figure 9.3*).

Potato starch often accompanies rusk in sausages, whereas corn starch is used more in canned meats, especially luncheon meat where chemically modified corn starch is a common ingredient. The starch in any canned product is always in an advanced state of gelatinization and is best seen when cryosections (10–12 μm) are stained with iodine vapour. For other methods of demonstrating starch see Chapters 6 and 8.

9.8   Experiments in sausage microscopy using simple equipment

Although a serious study of meat products involves the preparation of sections (preferably cryosections), some work is possible by using only the simplest of equipment and readily available materials. With coarse-chopped sausage, the obvious meat fragments can be isolated and examined as described earlier (see Section 9.4) but, in the simplified experiments which follow, advantage is taken of the finely comminuted nature of many mass-produced sausages. The finely divided structure of these products enables thin smears to be prepared in which proteins, fat and starch may be demonstrated. The sausage skin, or casing as it is called in the meat trade, also makes a suitable specimen for microscopical examination.

For microscopists who have difficulty in obtaining biological stains and reagents, food colours and Tincture of Iodine BP are suggested as alternatives to the more usual stains. Food colours are inexpensive, safe to use and available at most grocers and supermarkets. Many food colours contain familiar biological stains. The name of the dye or its additive number (E number) will be found on the label. The main disadvantage for their more general use is that the concentration of dye is unknown.

9.8.1   Materials needed

- Supermarket fresh pork or beef finely minced sausages.
- Food colours: pink, this usually consists of erythrosine (CI Acid Red 95) E number 127; green, usually consists of Green S (CI Acid Green 50) E number E142 often mixed with tartrazine (CI Acid Yellow 23) E number 102; other food colours may be used but these pink and green colours have been found to give consistently good results. Working solutions of both of these food colours are prepared by diluting 10 ml of colour to 25 ml with distilled water. The diluted solution keeps well.
- Iodine solution: Tincture of Iodine BP (obtainable from pharmacists). A working solution of iodine is prepared by diluting 1 ml of Tincture of Iodine to 20 ml with water just before use. The diluted solution will keep a few days if stored in the dark.

9.8.2   Apparatus

The basic equipment needed is a microscope which has a 10× objective (i.e. magnification of 60–100×) and a good light source. Optional, but desirable, are two polarizing filters (polars) but these require the microscope to have a fairly powerful light source whose intensity can be varied.

In addition to microscope slides and coverglasses, dropper bottles to hold the diluted stains and a plate and saucer or large watch glass are needed. Other tools required include scissors and a small spatula or paintbrush.

9.8.3 Experimental technique

Sausage skin. Unlike butchers' sausages which have skins made from natural gut, supermarket sausage skins are manufactured from skin collagen and cellulose fibres. These are much easier to handle than gut casings.

1. Take a sausage from the refrigerator and, using scissors, slit the skin along the sausage so that it can be peeled off.
2. Wash well to remove fragments of sausage meat using two changes of tepid water containing a few drops of detergent such as washing-up liquid. Rinse in cold water.
3. Spread on a plate and cut out 1-cm squares. Place these in clean cold water.
4. Put about 5 ml of diluted colour in a watch glass or saucer, and, using a spatula or paintbrush, transfer a square of skin to the stain. Leave for 5 min. Rinse in two changes of 100 ml water.
5. Transfer to a microscope slide, cover with water, and apply a coverglass. Blot away any excess water and examine with the 4× and 10× objectives using brightfield illumination and also crossed polars if these are available.

The results obtained are shown in *Figure 9.7.* Cellulose fibres in the skin are unstained but appear as thick curved fibres which are easy to see. Between crossed polars the fibres appear very bright and show polarization colours. The protein matrix surrounding the fibres stains pink or green (general background). The collagen fibres within the matrix are the thin straight fibres stained darker than the matrix; they are longer than the cellulose fibres but are not so easy to see. Make good use of the fine focus to trace the path of the longest fibres through the casing thickness.

Sausage meat: proteins and fat.
1. Make a smear by placing a 5-mm-diameter 'blob' of sausage meat towards one end of a microscope slide. With another slide firmly drag the blob into a thin smear. The smear must be thin enough (at least in places) to transmit light.
2. Place the smear on a plate or saucer and cover with the diluted food colour. Leave for 2 min.
3. Drain off the excess stain and rinse the slide in a large beaker of water. Wipe the underside of the slide and apply a coverglass. Invert and press on absorbent paper.
4. Examine with the 10× objective using brightfield illumination and crossed polars.

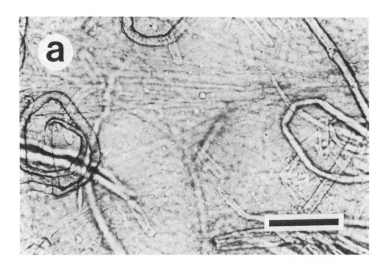

Figure 9.7: Wholemount of man-made sausage casing stained with green food colour. Bar = 100 μm. (a) Cellulose fibres and the fainter long straight collagen fibres seen with brightfield illumination. (b) The same field as (a) viewed with crossed polars showing strong birefringence of cellulose fibres.

The results are shown in *Figure 9.8*. Proteins stain pink or green; the fat and starch are unstained.

The long coloured strands are of connective tissue (collagen). Shorter fibres that show banding across the fibres are muscle fibres. The banding often looks like a thumb print. A network of stained collagen surrounds the unstained fat. Stained wheat protein has unstained starch embedded in it.

Crossed polars show brightly coloured connective tissue fibres, crystalline fat within rounded fat cells, and raw starch granules, mainly of wheat starch from the rusk but occasionally of added potato starch. The raw starches show a dark cross on a white background.

Sausage meat: starch.
1. Make a thin smear as described earlier. Remove any large fragments of meat or fat from the smear.
2. Add one or two drops of diluted iodine solution and leave for 1 min.
3. Apply a coverglass, invert and press on absorbent paper.
4. Remove excess smear from round the coverglass and examine the slide with brightfield illumination and between crossed polars.

The starch in sausage rusk contains gelatinized and raw wheat starch; raw potato starch may also be present. Unless the sausage is particularly greasy, most of the starch stains blue with iodine, the gelatinized starch staining most strongly. Staining tends to mask the brightness (birefringence) of the raw starch but that of the larger potato starch granules can normally be seen. Birefringent collagen fibres and crystalline fat are also visible (see *Figure 9.9*).

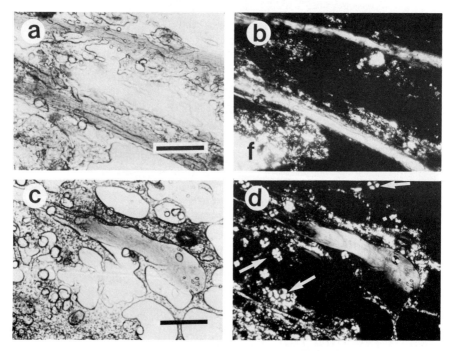

Figure 9.8: Protein in smears of sausage meat stained with pink food colour. Bar = 100 µm. (a) Long coloured strand of connective tissue (collagen) surrounded by unstained fat. (b) Same field as (a) viewed between crossed polars showing birefringent collagen and crystalline fat (f). (c) Shorter coloured fibres that show banding are of muscle fibre. The banding is often faint (as here); it is easier to see when the aperture (iris) diaphragm is closed down. (d) Same field as (c) viewed between crossed polars, birefringent muscle fibre shows banding, loose granules of raw starch are derived from rusk (arrows).

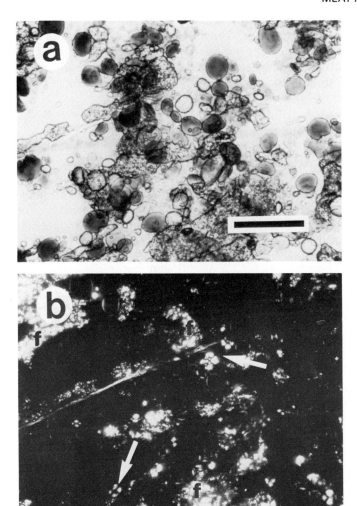

Figure 9.9: Starch in sausage smear stained with aqueous iodine. (a) Shows that the more swollen starch granules stain readily, the unstained granules are either raw or fat-covered. Bar = 100 μm. (b) Same field as (a) viewed between crossed polars shows biefringence of raw starch (arrows) and crystalline fat (f).

9.9   Fish and fish products

9.9.1   Fish muscle tissue

The flesh of bony fish consists of layers of muscle similar in structure to the striated muscle fibres present in meat, but much shorter, being only about 3 cm in length, and with both ends of the muscle fibres embedded in thin sheets of connective tissue. This connective tissue consists of a collagen which is more susceptible to moist heat than the collagen in meat. On cooking, the collagen quickly softens and the muscle fibres separate giving the cooked fish its characteristic flaky appearance (Howgate, 1979). Fish connective tissue has no elastin associated with it and, apart from the skin, the amount of collagen present in the edible part of fish is very small.

9.9.2 Fat in fish

The amount of fat present in the muscle tissue of white fish is small, being less than 1%. Although the flesh of oily fish contains much more fat (up to 20% in herring), this fat is not stored as adipose tissues as in meat animals, but is distributed throughout the flesh in a diffuse way.

The nature of fish flesh simplifies the microscopy of fish products because, unlike meat product manufacture, there is not the 'fat, rind, gristle and sinew' available as fish product ingredients.

9.9.3 Fish skin

Like animal skin, fish skin consists of interweaving bundles of collagen fibres but there is no elastin present and the structure is less ordered. Fish skin collagen cooks so readily that it can be left attached to fish muscle tissue during cooking and canning. Much more resistant to heat processing are the scales that cover the skin of bony fish.

9.9.4 Fish scales

Fish scales, in common with hairs and bristles on animals, contain keratin. The scales consist of a fibrous basal layer and a superficial layer often highly mineralized (Voehringer, 1979). With many commercial fish the scale patterns can be widely different, as shown by Essery (1922), but Essery also noted similarities in some closely related fish (cod, haddock, whiting and pollak). Even when the scale patterns appear similar, as in salmon species, a careful scrutiny of the scale pattern on entire scales can show differences. Five species of canned Pacific salmon have been identified from their scale patterns. Photomicrographs of the five different scales with a key to their identification is given in the AOAC Official Methods of Analysis (1990) paragraph 979.15.

Isolating entire fish scales from canned fish can be difficult and requires practice because the sterilizing process makes the scales very brittle. It is much easier to detach scales from fresh or lightly cooked fish and these can provide reference slides for use in the identification of fish scales present in fish products.

The preparation of fish scale mounts. The techniques for isolating and mounting salmon scales given in the AOAC method are recommended for preparing permanent mounts of scales from any fish. The important points are to keep the skin wet at all times and to use a thin-tipped spatula to remove the scales. The spatula is inserted between the scale and the skin and is then folded back to expose the whole scale which can then be removed. Fragments of scale pocket tend to adhere to the scale but these can usually be removed with a paintbrush, provided that the scale is kept wet. A stereomicroscope is useful for this operation. If the scale debris is difficult to remove, dilute sodium hydroxide can be used for cleaning the scale (Essery, 1922). The cleaned scales are mounted in glycerol jelly. The glycerol jelly used in the AOAC method is made by heating 5 g of gelatin with 60 ml of water and adding 40 ml (50.4 g) of glycerol and 0.5 g of phenol. This is cooled and stored at room temperature; some may be poured into a small Petri dish, where it sets quickly and is a convenient source of jelly for the mounting operation.

Mounting fish scales.
1. Place a small piece of glycerol jelly on a microscope slide and melt by placing the slide on a source of heat.
2. Pick up the scale with a microspatula or brush and remove excess water by touching gently on a tissue.
3. Place the scale in the melted jelly and cover with a coverglass.
4. Examine with the 4× objective or use a stereomicroscope to view the whole scale (*Figure 9.10*) and the 10× objective to see the detail of the scale pattern (*Figure 9.11*).

Permanent reference slides of fish scales provide an aid to the identification of fish scale fragments in fish products especially those made from crustaceans.

9.9.5 'Crustacean fish products'

'Crustacean fish products' include fish pastes made from lobster, crab and shrimp. As well as striated crustacean muscle and fat, these products often contain fragments of the hard shell which forms the protective covering of crustaceans. Unlike fish bones which consist of calcified collagen, crustacean shell is composed of chitin, an amino sugar polysaccharide. The chitin is calcified to varying extents depending on the species and the age of the shell. It follows that the presence of bones and scales in a crustacean fish paste indicates the use of bony fish as a paste ingredient. The bones and scales can be separated from a fish paste by treating the paste with alkali.

Separation of fish bone and scales from fish paste. Fish bones and scales can be isolated from a fish paste by the following method due to Essery (1922). Ten grammes of paste is mixed in a cream with 4% sodium hydroxide and then diluted to 150 ml with water. The mixture is warmed gently with frequent stirring for about 20 min. This separates the scales and bone which, along with any crustacean shell, sink to the bottom whilst the rest of the paste becomes flocculent and can be decanted. The bone,

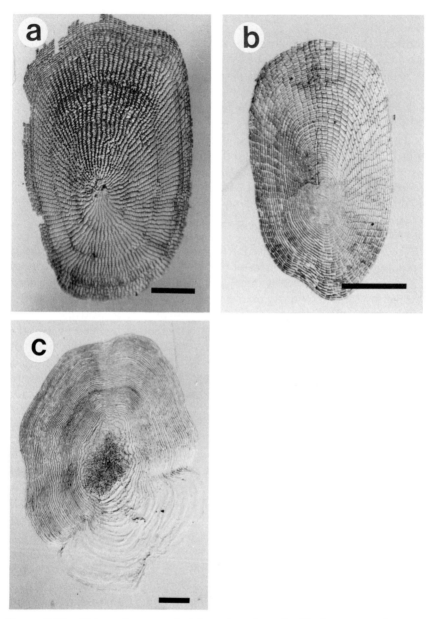

Figure 9.10: Fish scales mounted in water viewed with the stereomicroscope. Bars = 1 μm. (a) Haddock. (b) Cod. (c) Salmon.

scale and shell fragments are then washed thoroughly by decantation and examined in water using the 4× objective. The presence of fish bone and scale which, from their shape, are easy to identify indicates the presence of 'non-crustacean fish' in the paste (see *Figures 9.11* and *9.12*).

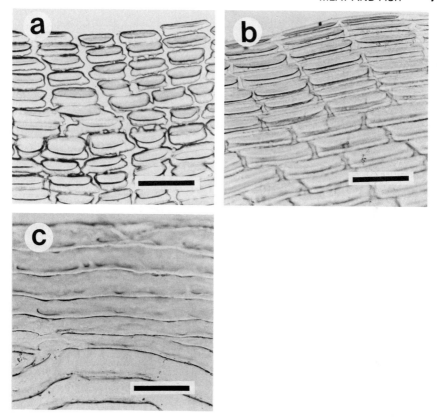

Figure 9.11: Fish scale detail; the scales illustrated in *Figure 9.10* viewed with 10× objective. Bars = 10 µm. (a) Haddock. (b) Cod. (c) Salmon.

Figure 9.12: Water mount of fish bone isolated from crab paste. Bar = 100 µm.

References

Association of Official Analytical Chemists (1990) *Official methods of the AOAC,* 15th edn. Association of Analytical Chemists, Washington, DC.

Bailey AJ. (1972) The basis of meat texture. *J. Sci. Food Agric.* **23**, 995–1007.

Essery RE. (1922) The value of fish scales as a means of identification of the fish used in manufactured products. *Analyst* **47**, 163–165.

Howgate P. (1979) Fish. In *Food Microscopy* (ed. JG Vaughan). Academic Press, London, pp. 343–389.

Lawrie RA. (1991) *Meat Science,* 5th edn. Pergamon Press, Oxford.

Voehringer H. (1979) Animal feeds – animal constituents. In *Food Microscopy* (ed. JG Vaughan). Academic Press, London, p. 428.

Further reading

Bailey AJ, Light ND. (1989) *Connective Tissue in Meat and Meat Products.* Elsevier Applied Science, London.

Burgess GHO, McLachlan T, Tatterson IN, Windsor ML. (1970) A new approach to the analysis of fish cakes. *Analyst* **95**, 471–475.

Campbell AM, Penfield MP, Griswold RM. (1980) *The Experimental Study of Food,* 2nd edn. Constable, London.

Lewis DF. (1979) Meat products. In *Food Microscopy* (ed. JG Vaughan). Academic Press, London, pp. 233–270.

Voyle CA. (1979) Meat. In *Food Microscopy* (ed. JG Vaughan). Academic Press, London, pp. 193–231.

10 Vegetable Proteins

10.1 Introduction

The richest source of protein in higher plants is embryonic tissue, that is seeds such as peas, beans, nuts and cereals. These all provide substantial amounts of dietary protein, but it is only from the soya bean and, to a lesser extent, wheat flour that high protein concentrates are made in any quantity. More recently, a protein-rich mycoprotein (Quorn) developed from the mould *Fusarium graminearium* has become widely available as a food ingredient.

10.2 The soya bean and soya products

The soya bean is an important commercial oil seed and its structure is well documented (Vaughan, 1970; Wallis, 1913; Winton and Winton, 1932). The bean is composed mainly of two cotyledons within a seed coat (hull) which is similar to, but distinct from, that of other pulses (Wallis, 1965). The cotyledons differ from other pulses in that they contain protein and oil but virtually no starch. The protein is in the form of aleurone grains (protein bodies) which are 4–18 µm in size (Vaughan, 1970). These are surrounded by minute (0.2–0.5 µm) membrane oil-bounded bodies (Smith, 1979) which, in the light microscope, appear as an oily matrix.

The structure of the seed coat is of interest because, although much of it is removed during processing, some adheres to the cotyledons and finds its way into food products where its presence helps in the identification of soya. In contrast to other legumes, it contains a distinct layer of aleurone cells which contain oil droplets and minute aleurone grains (Vaughan, 1970) but its most characteristic features are the closely packed rectangular palisade cells and their underlying hourglass cells (*Figure 10.1*).

Both types of cells are very resistant to heat and pressure and their strong birefringence survives heat processing, making them easy to identify in canned and extruder-cooked products.

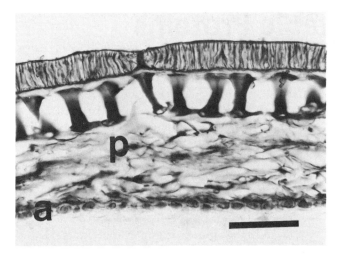

Figure 10.1:   Cryosection (20 μm) of soya bean hull stained with toluidine blue mountant showing outer palisade cells, hourglass cells, compressed parenchyma (p) and aleurone layer (a). Bar = 100 μm.

The palisade cells which have a radial diameter of about 50 μm form a double layer at the seed scar (hilum), this being the only area in the seed coat lacking hourglass cells. The hourglass cells are thick-walled lignified cells in contact with one another only at the top and base. They vary in height from 40 to 120 μm, the largest cells being near to the hilum. During processing, the palisade and hourglass cells tend to separate so that individual hourglass cells and small groups of palisade cells are found in soya-containing products.

10.2.1   Processed soya beans

Full-fat soya grits and flours. In the preparation of full-fat soya grits and flour, the beans are dried, cracked and most of the hulls removed by an air blast. The dehulled cotyledons are heat treated to destroy the trypsin inhibitors they contain and, after cooling, they are ground. A coarse grind yields a full-fat soya grit suitable for use as a nut extender in almond and other nut pastes. A fine grind results in a full-fat flour which is used extensively as an improver in commercial bread manufacture.

Defatted soya grits and flours. In the extraction of soya oil, the dried cracked beans are flaked and the oil removed using hexane. The defatted flakes may be ground to form defatted grits and flours which contain 52–55% protein, or the flakes can be treated with aqueous alcohol to remove soluble carbohydrates and yield soya concentrate which contains 70% protein. Grits and flours containing both levels of protein are widely used in comminuted meat products.

10.2.2 Microstructure of soya grits and flours

Apart from the lack of oil, defatted soya grits and flours show features similar to full-fat flours, for example all consist of cotyledon cells and are likely to contain traces of hull. The main difference between defatted and full-fat products lies in the additional heat treatment often given to defatted flakes to improve protein digestibility. Grits and flours which have received minimum heat treatment show separate protein bodies but where heat treatment has been more severe (as in toasted flakes) the protein bodies tend to fuse giving a single mass of protein in each cell (see *Figure 10.2*). Technically the flakes are graded into High Solubility Meals (white flakes) where the protein is undegraded, and Medium and Low Solubility Meals, the latter being golden brown toasted flakes largely used as animal feed.

High Solubility soya flakes are used to make soya protein isolate, a spray-dried product containing 95–98% protein which is used widely as an emulsifier with gel-forming properties.

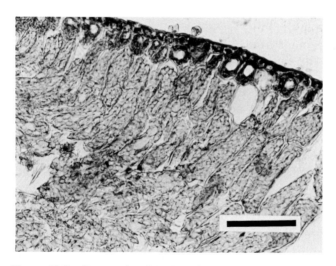

Figure 10.2: Cryosection (10 μm) of soya grit stained with toluidine blue mountant showing cotyledon cells containing partially fused protein bodies. Bar = 100 μm.

10.2.3 Microstructure of soya protein isolate

Spray-dried soya isolate consists essentially of hollow spheres of protein, but in the spray-drying operation some droplets are deformed and others partly coalesce before drying is complete (*Figure 10.3*). Although the particles are small enough for wholemounts to be made, sections are needed to show the structure of the isolate. During spray-drying, the protein in some particles is denatured, so that the isolate does not disperse completely when it is used in meat and vegetable protein products and it can often be seen in cryosections.

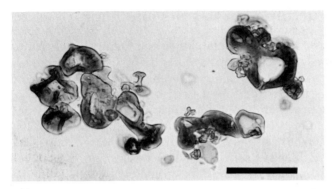

Figure 10.3:   Cryosection (10 μm) of spray-dried soya isolate stained with toluidine blue mountant showing. Note fusion of protein particles that occurred during drying. Bar = 50 μm.

10.2.4 Textured vegetable protein (TVP)

Textured vegetable protein (TVP), or texturized soya protein (TSP) as it is sometimes called, is made from defatted soya grits or soya concentrate by extrusion cooking with a small amount of water. During this continuous pressure cooking operation, most of the protein leaves the cotyledon cells and forms a continuous matrix which entraps the collapsed cotyledon cells and any seed coat particles present in the raw material. Aeration takes place as the product leaves the extruder when the sudden release of pressure produces cavities in the extrudate as water present in the dough flashes off as steam. The cavities extend in the direction of extension and can be so elongated that the product appears to have a fibrous structure (*Figure 10.4*). This can give the product a 'meaty' appearance, especially when it is seen in a hydrated state so that it can look like mince meat. By varying the conditions of extrusion, products varying from open-textured snack foods, such as 'bacon bits', to dense burger granules can be manufactured.

10.2.5 Microstructure of TVP

Some idea of the 'fibrous' nature of most TVP can be gained by examination of the hydrated material using a low power stereomicroscope (see Chapter 3) or by teasing the pseudo-fibres apart and making an aqueous mount of these. Sections of TVP are much more informative. Cryosections of TVP are easy to prepare (see Chapter 5) because the method of quenching is much less critical than it is for meat products. Cryosections of 10–12 μm cut along the axis of the fibres show the structure to the best advantage. These can be mounted in the toluidine blue stain mountant and viewed in brightfield and between crossed polars. This shows a purple-stained proteinaceous fibrous matrix enclosing pink-stained collapsed cotyledon cells and fragments of seed coat. The seed coat is less well stained but the palisade and hourglass cells are unmistakable and these show brilliant birefringence (*Figures 10.1* and *10.5*).

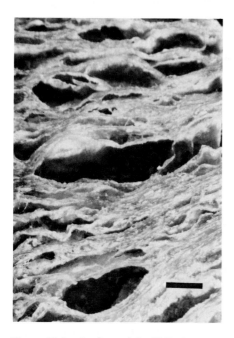

Figure 10.4:   Surface of dry TVP viewed with stereomicroscope. Bar = 1 mm.

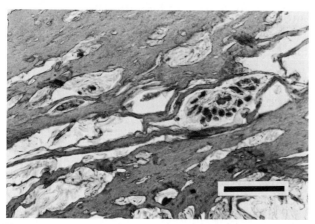

Figure 10.5:   Cryosection (12 μm) of TVP stained with toluidine blue mountant. Note entrapped cotyledon particle containing fused protein bodies. Bar = 100 μm.

10.3 Wheat protein

Wheat protein powder (gluten) is produced from wheat flour by making a flour–water dough and washing out most of the starch. The washed gluten is dried and ground and marketed as the so-called 'vital gluten'. The dry product contains 75–80% protein, 10–15% starch, about 5% fat and traces of wheat bran. Most gluten powder is used to fortify weak bread flours and to raise the protein level in low-calorie starch-reduced breads and biscuits. It is also used in high protein breakfast cereals and, like soya protein, as a meat extender and substitute. Canned gluten chunks were marketed as 'vegetarian steaks' long before vegetarian products based on extruded soya protein were introduced. Some of these are still manufactured but the trend in modern canned vegetarian products is to combine gluten and soya protein. This makes good nutritional sense because the essential amino acid profiles of soya and wheat protein are complementary. Soya protein has a high lysine content which wheat protein lacks, whereas wheat protein is high in the sulphur-containing amino acids which soya lacks. Gluten and soya protein are combined to make both canned and frozen vegetarian sausages (see colour plate frontispiece).

10.3.1 Microstructure of wheat gluten powder

The size and shape of gluten powder particles can be seen in an oil mount. The dry powder is difficult to stain unless it is first hydrated.

1. Place a small amount of powder (only 1–2 mg) on a slide, tap out to separate the particles and add 1 drop of water.
2. Leave for 1 min and then add 1 drop of diluted iodine, leave for a further minute and then position a coverglass over the preparation.

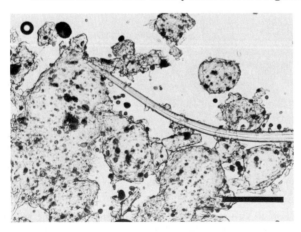

Figure 10.6: Wholemount of wheat gluten powder stained with aqueous iodine. Note small size of starch granules and the trichome cell. Bar = 100 μm.

3. Invert the slide over blotting paper and press down firmly. This will distort the swollen particles but enable the blue-stained starch granules to be seen more clearly. Note the size of the granules in the gluten; these tend to be the smallest granules from the original flour because the larger granules are washed out more easily when the gluten is prepared (*Figure 10.6*).

4. Make a second hydrated gluten preparation and add 1 drop of 0.1% aqueous toluidine blue, leave for 1 min and blot as before. This will stain the protein a light blue but leave the starch granules unstained so that the birefringence of even the smallest granules can be seen when the slide is viewed between crossed polars. The bran particles which are stained a darker blue are also birefringent.

A comparison of the oil-mounted gluten particles with either of the stained mounts will show how much the structure of the gluten powder has been altered in the preparation of the wet mounts. A truer and more informative picture of the gluten is obtained from 10–12 μm cryosections of the powder which can be stained with iodine vapour. A very light iodine treatment only colours the starch granules gelatinized during the drying of the gluten. More iodine vapour stains all the starch blue and the protein a pale yellow. Heavier staining colours the protein a golden yellow with the starch a blue–black colour. Mount each preparation in light mineral oil and examine with brightfield illumination and between crossed polars.

Traces of wheat bran can often be seen in gluten preparations, especially fragments of aleurone layer and broken trichome cells (*Figure 10.6*). The aleurone forms a single layer in the bran consisting of close-packed thick-walled cells which contain protein and fat. The trichomes form a tuft of 'hairs', known as the beard, which is found at the apex of the wheat grain. Trichomes are thick-walled cells up to 1 mm long and broad at the base (12–20 μm) tapering to a point. Fragments of them are easily recognized by their shape and strong birefringence.

10.4 Mycoprotein

The mycoprotein which is now commercially available and marketed as 'Quorn' is a frozen product based on the mycelia (hyphae) of the mould *F. graminearium*. It has a naturally fibrous structure, so that Quorn chunks and mince simulate meat. Quorn is now sold as an ingredient for home cooking and forms part of several convenience foods, including pies and cook/chill dishes. The mould is grown in a continuous fermenter being fed by glucose syrup plus a nitrogen source and minerals. After treatment to remove excess nucleic acids, the filter mat is rolled and folded so that the mould filaments all face the same way. It is then tumbled with egg albumen and vegetable extract, steamed to set the egg protein, and chopped into cubes or mince which is packed and frozen.

On a dry basis, Quorn contains 55% protein, 13% fat and 22% fibre.

10.4.1 Microstructure of Quorn

Growing conditions affect the thickness and length of the mould filaments but they are usually 1–2.5 µm thick and a few millimetres long. The fibres can be teased apart and stained but their fineness and the egg albumen which glues them together make it difficult to produce satisfactory wholemounts (*Figure 10.7*).

Quorn sections easily and 4–6 µm cryosections can be mounted in either the trypan blue or the toluidine blue stain mountant (*Figure 10.8*).

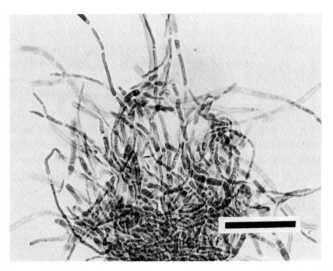

Figure 10.7: Wholemount of mycoprotein filaments stained with trypan blue mountant. Note that the mould is septate. Bar = 50 µm.

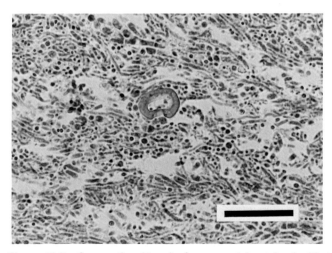

Figure 10.8: Cryosection (4 µm) of mycoprotein stained with trypan blue mountant. Note particle of undispersed egg albumen. Bar = 50 µm.

References

Smith CG. (1979) Oil seeds. In *Food Microscopy* (ed. JG Vaughan). Academic Press, London, pp. 35–71.
Vaughan JG. (1970) *The Structure and Utilization of Oil Seeds.* Chapman and Hall, London.
Wallis TE. (1913) The structure of the soya bean. *Pharmacol. J.* **91**, 120–123.
Wallis TE. (1965) *Analytical Microscopy,* 3rd edn. J. and A. Churchill, London.
Winton AL, Winton KB. (1932) *The Structure and Composition of Foods*, Vol. 1. John Wiley and Sons, New York.

Further reading

British Food Standards Committee (1974) *Report on Novel Protein Food.* FSC/REP/62. HMSO, London.
Flint FO. (1979) Novel protein foods. In *Food Microscopy* (ed. JG Vaughan). Academic Press, London, pp. 531–549.
Flint FO. Johnson RFP. (1979) Histochemical identification of commercial wheat gluten. *Analyst* **104**, 1135–1137.

11 The Howard Mould Count of Tomato Products

11.1 Introduction

As well as their use as a fresh vegetable, tomatoes head the list of vegetables that are canned, and large amounts are concentrated to yield tomato paste, purée and powder. Tomato purée and tomato paste (which has a higher solids content) are used in the manufacture of ketchups, sauces, including those for canned beans and spaghetti, and canned soups. Tomato powder is an ingredient of dried products such as dry soups and casserole mixes.

All tomato products contain mould filaments (hyphae) because the tomato is very susceptible to mould infection. Moulds infect the growing tomato and develop rapidly in ripe fruit picked and held prior to processing. In itself the presence of mould filaments is not harmful, but when a tomato product contains an abundance of hyphae it is a clear indication that unsound fruit has been used, and/or that it was processed unhygienically.

A tomato product made from unsound fruit or one processed with unclean machinery cannot be regarded as wholesome, and often it carries an off-flavour. However, tomato products invariably contain some mould filaments so a problem arises in deciding acceptable limits for the amount of mould present, and how this is to be assessed.

11.2 The Howard mould count and its limitations

The current method for the assessment of moulds in tomato products is the Howard mould count. This is an empirical method that was introduced in 1911 by B.J. Howard as a means of assessing the quality of commercial tomato ketchup in the USA. Howard reasoned that tomato ketchups made from bad tomatoes would contain mould filaments and he showed a relationship between the number of mould filaments in a ketchup and the proportion of rotten fruit used in its making (Howard, 1911). Howard used an ordinary microscope slide to hold diluted ketchup and, using the 10× objective, he noted the percentage of microscope fields that contained mould filaments. This percentage became known as the Howard mould count. Later the method was standardized and the Mould Count Chamber, a cell with a depth of exactly 0.1 mm, came into use.

The Howard method has been much criticized. Rot caused by different moulds contains different amounts of mould hyphae and breaks up in different ways. A low count does not necessarily indicate the use of sound fruit but a high count always shows the use of mouldy fruit and/or dirty machinery. There is also the risk of microscopist error. The ability to discover and identify mould filaments in the presence of comminuted fruit tissues, some of which may have a 'mould-like' appearance, requires practice and is best learned by personal instruction from an experienced mould counter.

It has been thought that the estimation of chitin, which is a characteristic constituent of mould cell walls, could replace the Howard mould count. However, chitin levels have been found to vary between mould species and with the age of the mould and there is also the possibility that insect fragments which contain chitin could be present (Jarvis and Williams, 1987).

Howard counts can only be done with any degree of confidence when the microscopist is familiar with the appearance of the different tissues present in the tomato, as well as the different forms of mould filament which may be present. A study of tomato histology should take place before any attempt is made to prepare a sample for the Howard cell (Anon, 1968).

11.3   Tomato histology

Place a fresh, ripe tomato in a beaker and pour boiling water over it. Leave for 1 min and transfer to cold water. From this tomato with its loosened skin, isolate the following tissues as indicated, mount in water and examine with the 10× objective. View each preparation with brightfield illumination and with crossed polars.

11.3.1 Skin cells

Remove a small portion of the skin from the tomato, place on a microscope slide, and vigorously scrape away the adhering flesh cells using a spatula. When all the flesh cells are removed, that is when the skin appears yellow, rinse in water and mount.

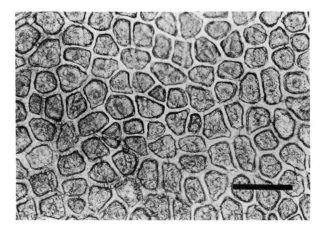

Figure 11.1:   Skin cells of tomato. Bar = 100 μm.

Note the closely packed thick-walled golden yellow cells which are birefringent (*Figure 11.1*).

11.3.2 Flesh cells

Sample these from any part of the tomato flesh, taking only a tiny piece of the fleshy tissue. Place this between two microscope slides, compress the slides and shear them apart. Add a drop of water to each slide and mount.

The shearing action will have broken some of the cells so that they appear as they do in tomato products. All flesh cells have very thin cellulose walls which are so transparent they have been likened to 'cellophane footballs' (Anon, 1968). The cell contents include tiny particles of red-coloured lycopene pigment and, very occasionally, small starch granules; both of these are birefringent. Note especially the folds present in some cells because these elongated creases can look like mould filaments. Between crossed polars the creases show birefringence whereas mould filaments do not (*Figure 11.2*).

11.3.3 Fibrovascular tissue

The tomato has a system of fibrous 'threads' which serve to carry water and nutrients around the fruit. Isolate this fibrovascular material by cutting the tomato in half and teasing out one of the white threads found near the stem end of the fruit. Place the thread on the centre of a microscope slide containing a drop of water and put another slide on top and at right angles to the first slide. Vigorously shear the trapped tissue between the two slides until it appears translucent. Separate the slides and add a drop of water and a coverglass to one of them. Blot firmly to remove excess water and examine.

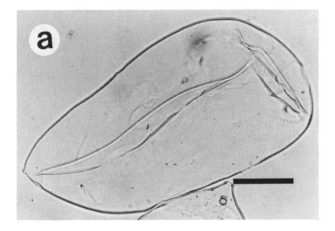

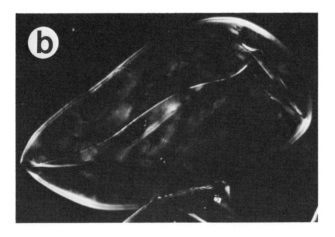

Figure 11.2: Tomato flesh cell. Bar = 100 µm. (a) Viewed with brightfield illumination, note creases in cell wall. (b) Viewed between crossed polars showing birefringence of cell outline and creases.

The shearing action should have broken and 'unravelled' some of the coiled xylem vessels showing them as they frequently appear in tomato products where small lengths of uncoiled xylem may simulate mould filaments. Xylem vessels may be distinguished by their birefringence (*Figure 11.3*).

11.3.4 Seed 'hairs'

Seed hairs are another mould 'look-alike' so are well worth careful study. Take a tomato seed from the halved tomato and place on a microscope slide. Hold it firmly with one mounted needle and with another remove

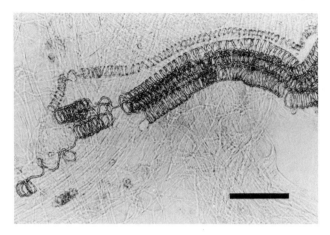

Figure 11.3:   Fibrovascular tissue from tomato flesh. Bar = 100 μm.

the jelly-like coating which surrounds the seed. A dense coat of false hairs covering the seed is now just visible as a furry 'halo' round the seed. Pin down the seed with one needle and with another detach a small fragment of the seed coat and transfer it to another slide. Mount and examine. Each 'hair' looks like a long tapering icicle. It is a single elongated epidermal cell which may be up to 500 μm in length. Fragments of seed hairs are frequently found in tomato products; they can be distinguished from mould filaments by their tapering nature and birefringence *(Figure 11.4)*.

11.4 Mould histology

Either obtain tomatoes with obvious mould infection or store one or two tomatoes in a moist atmosphere for a week or two until mould growth is apparent. If this is not possible, an examination of any available moulds growing on food will have to suffice.

Remove a little of the mould growth and place on a microscope slide with a drop of water. If the mould is difficult to wet, first treat it with a drop of alcohol, allow to dry and then add water. Apply a coverglass and firmly blot any excess water. Examine the slide using both the 10× and 20× objectives with brightfield illumination and with crossed polars.

Note that although any adhering tomato tissue is birefringent the mould filaments are not. The observation of any fruiting heads present in the mould is not relevant to mould counting, these are almost never seen in tomato products because any present on the raw tomatoes are removed when the fruit is washed before processing. By using the 10× and 20× objectives some or all of the following characteristics of moulds will be seen *(Figure 11.5)*.

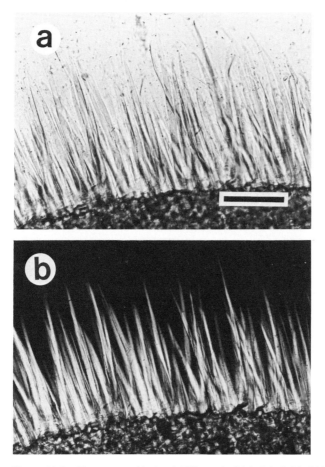

Figure 11.4:   Tomato seed hairs. (a) Viewed with brightfield. (b) Viewed between crossed polars. Bar = 100 µm.

11.4.1 Parallel filament walls

Although they appear flat under the microscope, mould filaments are tubular, and for any one filament the tube diameter is uniform so that the filament walls are parallel. This is an important diagnostic feature although there are exceptions. For example, *Geotrichum* (*Oospora*) the machinery mould and some *Mucor* species have filaments with tapered ends; their lack of birefringence confirms them as moulds.

11.4.2 Branching of filaments

Many moulds show an abundance of branching and this is a reliable characteristic for mould identification.

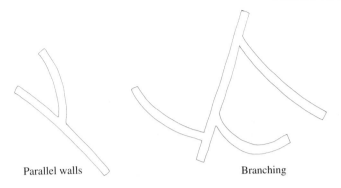

Parallel walls   Branching

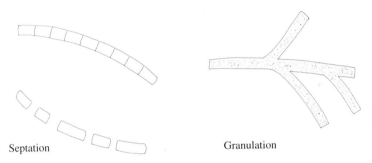

Septation   Granulation

Figure 11.5:   Four features which characterize moulds. Only filaments present in tomato products which show one or more of these features may be classed as mould hyphae (after Gould,1983).

11.4.3 Septation

Some, but not all, mould filaments are divided into segments by cross walls. These provide another reliable characteristic for mould identification.

11.4.4 Granulation

Growing moulds contain protoplasm within the filaments which often present a granular or stippled appearance. This is most easily seen in mould filaments of larger diameter, for example some *Mucor* spp., but is less obvious in slender mould filaments.

11.5 Practical mould counting

The Howard cell used in the UK and marketed by Graticules Ltd is of a slightly different construction to that described in the official Methods of the Association of Official Analytical Chemists (AOAC, 1990). The UK version consists of a thick glass slide (75 mm × 35 mm) which has a central flat circle 19 mm in diameter which is precisely 0.1 mm lower than the main slide. Surrounding this circle is a moat which is about 3 mm deep. When the optically flat Howard coverglass is laid on the slide, a cell exactly 0.1 mm deep is formed (*Figure 11.6*).

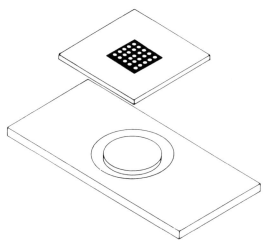

Figure 11.6:   Howard cell and coverglass containing 25 calibrated fields. The surface of the central 'plateau' is 0.1 mm lower than that of the main slide; the 'plateau' is surrounded by a groove forming a 'moat'.

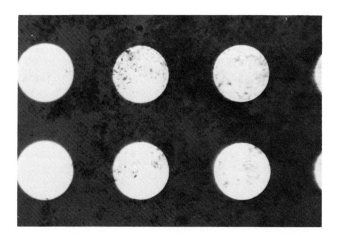

Figure 11.7:   Howard cell and coverglass with a mount of tomato ketchup as seen with the 4× objective.

In the Howard method, 25 evenly spaced fields are scanned at a time. Each field must cover 1.5 mm^2 which is given when the field diameter is exactly 1.382 mm. This can be achieved by mounting a coloured cellophane grid containing 25 holes with this diameter on the upper surface of the Howard coverglass using a mountant with a refractive index of 1.53 and a very thin coverglass. Coverglasses which incorporate similar grids are available from Graticules Ltd (*Figure 11.7*).

To ensure that the cell depth of 0.1 mm is not exceeded both Howard cell and coverglass must be very clean. Wash in water, rinse in alcohol and polish with a clean cloth. Place the coverglass in position and press the ends of it gently. Newton's 'rainbow' rings should be seen between the cell and the coverglass. Do not press the central area of the coverglass as this may easily break it.

11.5.1 Preparing the sample for the Howard cell

1. Tomato juice and tomato sauce: use as it comes from the container.
2. Canned tomatoes: use the drained liquid.
3. Tomato ketchup (without added thickener): dilute 1:1 with 0.5% carboxymethyl cellulose.
4. Tomato ketchup containing thickener: dilute 1:1 with water.
5. Tomato paste and purée: dilute with water to make a mixture having a refractive index of 1.3448–1.3454 at 20°C. (This corresponds to a solids content of 8.0–8.4%). Before checking the refractive index, the diluted paste or purée must be well mixed. This is conveniently done by pouring from one container to another which avoids trapping air in the diluted product. Check the refractive index by filtering a portion, reject the first 1 ml and measure the refractive index of a drop of filtrate using a refractometer.
6. Tomato powder: add 17.0 g of powder to 150 ml of water contained in a high-speed blender. Blend for 30 sec, rub down the material adhering to the walls of the blender and rinse the walls with 50 ml of water. Blend for 1 min and add 2 drops of octan-2-ol to break the foam formed.

Methods of sample preparation for other tomato products including canned soup, the sauce from canned beans and the purée used as a packing medium for canned fish, are given in the 1990 AOAC methods paragraphs 945.91–945.93.

After appropriate dilution it is advisable to add 2–4 drops of octan-2-ol to each 100 ml of the count preparations. This will help to eliminate air bubbles on the counting slide.

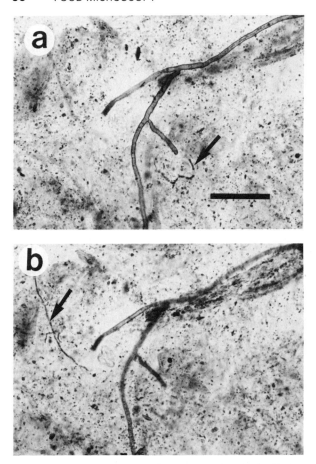

Figure 11.8:   A preparation of tomato ketchup in the Howard cell showing the need for focusing through the depth of the cell. (a) and (b) are of the same field seen at different depths. Note the disappearance of two fine mould filaments (arrows) as the focus is changed. Bar = 100 μm.

11.5.2 Mounting the sample

Remove the coverglass from the Howard cell and place the sample drop on the central disc. This is a critical operation and may require practice. The important points are:

1. To add just enough well-mixed sample to cover the central disc entirely with possibly a little in the 'moat' but not on the shoulder;
2   To avoid air bubbles;
3. To obtain an even distribution of insoluble material because mould fragments are often trapped in this.

Amongst the methods suggested for obtaining of homogenous spread (including the current AOAC method) the technique of Williams (1968) is

recommended. This uses a straight-sided tube of 3–3.5 mm bore to de-
liver a single large drop of well-mixed sample on the cell disc. The drop is
spread evenly with a needle, carefully avoiding scratching the disc, and
the Howard coverglass is lowered gently on to the slide. Before counting,
check that the cell contains an even distribution of insoluble material by
viewing it over a light source or by the use of a low power stereomicroscope.

11.5.3 Counting technique

Using the 10× objective and 10× eyepiece, scan each of the 25 fields that
the grid provides. Optimum illumination of the microscope is crucial, the
light intensity must be sufficient to penetrate insoluble material so that
mould can be seen, but not so bright that fine mould filaments are con-
cealed. Scan each field, if it clearly shows mould filaments record as positive and pass on to the next field. If the field appears negative or nearly
so, a detailed search is needed. This involves focusing up and down the
depth of the cell using the fine adjustment to bring into view mould filaments that may be at different depths in the 100 µm cell (see *Figure
11.8*). Varying the intensity of the light may be helpful at this stage. Watch
out for small fragments of mould because if the total length of not more
than three of these is greater than one sixth of the field diameter then
that field is positive. When in doubt about the identity of a mould, bring
it to the centre of the field and view with the 20× objective using brightfield
illumination and also crossed polars with increased illumination. If the
suspect material is tomato tissue it will show birefringence.

Examine three mounts (75 fields). If the separate counts of 25 fields
vary by more than two, prepare further mounts until agreement is
reached. The result is expressed as percentage positive fields.

The AOAC method requires a minimum of 50 fields to be examined
but it is advisable to count more unless the sample has an especially low
count. A suggested method of recording results is to draw a score card of
25 dashes corresponding to the coverglass grid. Convert a dash (–) to +
for each positive field as in (a) and (b).

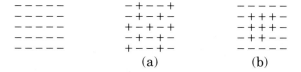

(a)   (b)

If the final score shows a marked concentration of + signs as shown in
(b), this indicates an unevenly prepared slide and this count should be
rejected. This is an error which is easily reproduced giving consistently
low results, for example (b) = 32% with (a) showing 44%.

Despite criticisms, the Howard mould count is still the only generally
accepted method for assessing the quality of tomato products. Where
mould counting is done on a regular basis, mould count record forms may
be kept (Gould, 1983). These are similar to the score card described above

but each of the 25 fields is represented by a circle divided into four. All the mould filaments seen are sketched in the appropriate field to show their relative size and position. Such forms are useful in training because they allow a count to be checked by another observer.

Over the years, the Howard method has been adapted for other fruit-based products, including apple batter, frozen and canned drupelet berries, citrus and pineapple juices, cranberry sauce and puréed fruits and methods for these are described in the AOAC Official Methods (1990).

11.6 Mould count tolerances

The first Federal Food and Drug Administration limit for the Howard mould count of tomato purée was 66% positive fields set in 1916. Since then the tolerance level has been progressively reduced until, by 1941, it had reached a level of 40% for tomato purée and 20% for tomato juice (Anon, 1965). Different maxima for mould counts apply in different countries (Williams, 1968) but there is no legislation for tomato products in the UK. In its absence, the Association of Public Analysts has adopted a maximum limit of 50% positive fields for tomato purée and 25% for tomato juice and there have been prosecutions when these figures were greatly exceeded (Kirk and Sawyer, 1991).

An important aspect of the mould count is that it is a quantitative measure of tomato quality. Manufacturers using tomato purée as an ingredient of sauces and canned products can include their own mould count limits in the specifications they set for their suppliers.

References

Association of Official Analytical Chemists (1990) *Official Methods of the AOAC*, 15th edn. Association of Analytical Chemists, Washington, DC.

Anon (1968) *Mold Counting of Tomato Products*. Continental Can Co. Technology Center, Chicago. (Out of print. Useful practical guide to the Howard mould count.)

Gould WA. (1983) *Tomato Production, Processing and Quality Evaluation*, 2nd edn. Avi Publishing Company, Westport, CT, p. 323.

Howard BJ. (1911) Tomato ketchup under the microscope with practical suggestions to insure a cleanly product. *US Department of Agriculture Bureau of Chemistry*, Circular No 68.

Jarvis B, Williams AP. (1987) Methods for detecting fungi in foods and beverages. In *Food and Beverage Mycology*, 2nd edn (ed. LR Beuchart). Van Nostrand Reinhold, New York, pp. 599–636.

Kirk RS, Sawyer R. (1991) *Pearson's Composition and Analysis of Foods*, 9th edn. Longman Scientific and Technical, Harlow, p. 255.

Williams HA. (1968) The detection of rot in tomato products. *J. Assoc. Public Analysts* **6**, 69–84.

12 Food Gums

12.1 Introduction

The trend towards low-calorie and convenience foods has meant a greatly increased use of food gums as thickeners, stabilizers and emulsifiers.

Food gums are high molecular weight, water-soluble carbohydrates (hydrocolloids) derived from natural sources. They include: pectin from fruit, carboxymethyl cellulose (CMC) from cellulose fibres, seed gums (guar and locust bean (carob)), gum exudates (acacia, ghatti, karaya and tragacanth), seaweed extracts (agar, carrageenan, Irish moss and alginates) the microbial polysaccharide xantham gum and pre-gelled starches (Whistler and BeMiller, 1993).

12.2 Reactions of gums with toluidine blue

Many gums are imported in a powdered form and there is often a need to confirm their identity. This is difficult to do by chemical methods but, by observing the microscopic appearance of different gum particles when these are treated with aqueous toluidine blue under neutral and acid conditions, it is possible to identify each of the above gums (Flint, 1990).

The effects of staining with aqueous toluidine blue are not empirical. They can be explained in terms of the physical and chemical properties of the gums. Powdered gums respond to water in different ways: gum acacia and pectin dissolve rapidly, others like tragacanth and karaya gum first absorb water and swell before eventually dispersing, whereas agar and gum ghatti absorb water relatively slowly. This affects the physical structure of the gums when they are treated with aqueous dyes. With the exception of pre-gelled starch, all gums contain acid groups which can bind the basic metachromatic dye toluidine blue, but different gums have different dye-binding properties. When stained with toluidine blue, the hydrated gum particles are coloured in different shades of blue, purple and pink according to the density of the bound dye molecules. When acidified, the carboxyl-containing gums no longer bind toluidine blue but

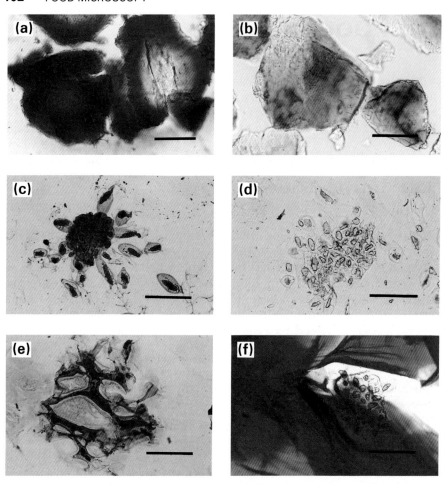

Figure 12.1: Food gum powders. Bars = 100 μm. (a) Agar stained with aqueous toluidine blue. (b) Agar stained with aqueous iodine. (c) Guar gum stained with aqueous toluidine blue. (d) Guar gum stained with iodine. (e) Locust bean gum (carob) stained with toluidine blue. (f) A mixture of carrageenan and guar gum stained with aqueous toluidine blue. Note lack of staining of guar cells (compare with (c)).

the sulphate-containing seaweed gums retain their anionic charge and staining properties even at pH 2 (*Figure 12.1*).

12.3 Reactions of gums with iodine

Iodine is needed to give a positive test for starch-based gums and it also serves as a confirmatory test for tragacanth, guar, carob, Irish moss, agar and carrageenan, all of which stain with iodine.

12.4 Birefringence of stained gums

The use of crossed polars to demonstrate the birefringence of stained gums is not strictly necessary for their identification but it can be revealing, and it involves little extra work. It enables the starch granules present in gum tragacanth to be seen in the toluidine blue mount and any ungelatinized starch present in a pre-gelled starch product to be recognized. The palisade cells from the seed coats of guar and locust bean are strongly birefringent as is the bark-derived debris present in many samples of the exudate gums. These can be seen in both the toluidine blue and the iodine mounts, being most abundant in technical grades of the gums.

12.5 The identification of commercial food gums (Flint, 1990)

It is useful to have a range of authentic gum samples both to serve as reference material for the identification of an unknown gum, and also to check the quality of the reagents used. All the gums permitted as food ingredients are available from Sigma Chemical Co. Ltd.

12.5.1 Reagents needed

Aqueous toluidine blue, 0.1%. It is important to use a dye sample of approximately 90% purity because staining is adversely affected by the presence of inorganic salts. Some samples of toluidine blue contain only 50% dye. These consist of the zinc chloride double salt of the dye and are not suitable. Toluidine blue certified by the Biological Stain Commission obtainable from Sigma Chemical Co. Ltd or Aldrich Chemical Company Ltd is recommended.

Prepare a 0.1% aqueous dye solution of toluidine blue by dissolving the appropriate amount (i.e. approximately 110 mg) of certified toluidine blue in 100 ml of distilled water.

Oxalic acid, 0.2 M. Prepare by dissolving 2.52 g of oxalic acid hydrate in distilled water and diluting to 100 ml.

Stock solution of iodine in aqueous potassium iodide. Prepare by dissolving 1 g of iodine in 100 ml of 2% aqueous potassium iodide. Store in an amber glass-stoppered bottle.

Working solution of iodine. Prepare by diluting 5 ml of stock iodine solution to 100 ml with distilled water just before use.

12.5.2 Toluidine staining of hydrated gums

1. Sprinkle 2–3 mg of powdered gum on a microscope slide.
2. Tap the slide to give an even layer of the powder about 1 cm in diameter.
3. Add 1 drop of distilled water, leave to hydrate for 1 min.
4. Add 1 drop of 0.1% toluidine blue, leave to stain for 1 min.
5. Place a coverglass over the stained gum, invert the slide and blot firmly on fluff-free blotting paper.
6. Examine the slide by brightfield illumination, and between crossed polars using a 10× objective.

All gums except pectin, gum arabic and pre-gelled starch are stained positively (see *Table 12.1* and the illustrations in Flint, 1990).

12.5.3 Method for pectin, gum arabic and pre-gelled starches

Heap about 3 mg of gum in the centre of a microscope slide, do not tap out or hydrate, but add 1 drop of 0.1% toluidine blue to the gum. Quickly position a coverglass, blot, and examine straight away as for the stained hydrated gum. Pectin and gum arabic will now give a positive test (see *Table 12.1*), pre-gelled starch remains uncoloured (see iodine staining for confirmation).

12.5.4 Toluidine blue staining of acidified gums

1. Sprinkle 2–3 mg of powdered gum on a microscope slide.
2. Tap the slide to give an even layer of the powder about 1 cm in diameter.
3. Add 1 or 2 drops of 0.2 M oxalic acid ensuring that all the gum particles on the slide are wetted. Leave for 1 min.
4. Add 1 drop 0.1% toluidine blue, leave for 1 min.
5. Position a coverglass, invert and blot firmly on fluff-free blotting paper.
6. Examine the slide by brightfield illumination, and between crossed polars using a 10× objective.

Only agar, Irish moss, carrageenan, alginic acid and sodium alginate stain positively (see *Table 12.1* and the illustrations in Flint, 1990).

12.5.5 Iodine staining of gums

1. Sprinkle 2–3 mg of powdered gum on a microscope slide.
2. Tap the slide to give an even layer.
3. Add 1 drop of freshly diluted iodine solution, leave for 1 min.
4. Apply a coverglass, invert and blot.
5. Examine the slide by brightfield illumination and between crossed polars.

Pre-gelled starches, tragacanth, guar, locust bean (carob) agar and Irish moss give well-defined staining. Refined carrageenan stains very weakly so that the colour is best seen by eye when the slide is viewed against a white background, semi-refined carrageenans resemble Irish moss. See Flint (1990) and *Table 12.2* for details.

Table 12.1: Microstructure of gums stained with toluidine blue

Gum	Appearance of hydrated gum stained with toluidine blue	Appearance of acidified gum stained with toluidine blue
Guar	Pink cells with dark purple contents	Colourless cells with pale blue contents
Locust (carob)	Pink 'foam-like' structure containing irregular dark purple bodies	Colourless 'foam-like' structure containing irregular pale blue bodies
Tragacanth	Pink, blue and purple swollen particles containing small intact starch granules (unstained) which show well-defined Maltese crosses when viewed between crossed polars	Unstained swollen particles starch granules prominent. Crossed polars show well-defined Maltese crosses
Karaya	Pink and dark magenta 'puffy' swollen granules (cumulus cloud shape), shows bright polarization colours	Unstained
Ghatti	Dull purple angular particles showing parallel striations, weakly birefringent	Unstained
Arabic (acacia)	Pale purple–pink angular particles 'melt and disperse', not birefringent	Unstained
Pectin	Deep magenta pink rounded particles, fast dispersing, contain tiny birefringent particles	Unstained
Agar	Dark magenta purple staining outer layers of angular particles, particle interiors unstained, brightly birefringent	Same staining as hydrated gum
Irish moss (crude powder)	Bright magenta dispersion showing darker stained 'folds'. Rounded groups of cells enclosed, shows bright polarization colours	Same staining as hydrated gum

Table 12.1: Microstructure of gums stained with toluidine blue, *continued*

Gum	Appearance of hydrated gum stained with toluidine blue	Appearance of acidified gum stained with toluidine blue
Carrageenan	Intensely stained magenta dispersion showing dark stained 'folds', shows bright polarization colours	Same staining as hydrated gum
Alginic acid	Swollen pink particles dispersed in a pink–purple matrix, particles and matrix show polarization colours	Swollen purple particles, no dispersed gum, particles birefringent (pink and blue)
Sodium alginates	Intensely stained pink and dark purple dispersion, bright polarization colours similar to carrageenan	Swollen purple particles, some samples show pink 'exudate' round particles but particles substantially undispersed, low level birefringence (mainly blue)
Xanthan	Intensely stained magenta dispersion showing dark stained 'folds' very similar to carrageenan, bright polarization colours	Unstained
Pre-gelatinized starch	Unstained, any intact starch granules present show birefringence	Unstained
Sodium CMC high, medium and low viscosity samples	All samples disperse well and stain intensely, either magenta or dark purple with dark stained folds, occasional fibre-shaped particle (well stained) occurs, bright polarization colours similar to carrageenan	Marked differences in degree of dispersion of pale stained fibres. Fibres almost invisible in brightfield can be seen between crossed polars

Table 12.2: Microstructure of gums stained with iodine reagent

Gum	Appearance of gum stained with iodine reagent
Guar	Colourless cells with yellow contents
Locust (carob)	Colourless structure with irregular yellow bodies
Tragacanth	Unstained swollen particles containing heavily stained blue–black starch granules (staining masks birefringence)
Karaya	Unstained
Ghatti	Unstained
Arabic (acacia)	Unstained
Pectin	Unstained
Agar	Patchy deep red–brown angular particles, birefringent
Irish moss (crude powder)	Unstained matrix enclosing strongly stained red–brown filled cells
Carrageenan (refined)	Viewed without microscope, slide appears pale purple–brown
Carrageenan (semi-refined)	Similar to Irish moss
Alginic acid	Unstained
Sodium alginates	Unstained
Xanthan	Unstained
Pre-gelatinized starch	Red stained swollen gum particles containing blue starch granules, staining masks birefringence of any ungelatinized granules
CMC	Unstained

12.6 Identification of powdered gum mixtures

Gums may be mixed for reasons of economy or because the gum mixture has synergistic properties, for example the addition of locust bean gum (carob) greatly improves the gelling power of carrageenan even though locust bean itself is non-gelling. The toluidine blue staining characteristics of individual gums are also affected when gums occur as mixtures. For example, when guar or locust bean gum is mixed with a rapidly dispersing gum, such as carrageenan or xanthan, the rapidly dispersing gum stains as expected but the seed gum is understained, appearing as if it had been pre-acidified, that is its cellular structure is visible but it shows barely any colour (*Figure 12.2*). Tragacanth gum is also 'deprived' of stain when mixed with the more rapidly dispersing gums but it can be recognized by its starch granule content which is best seen with the 20× objective.

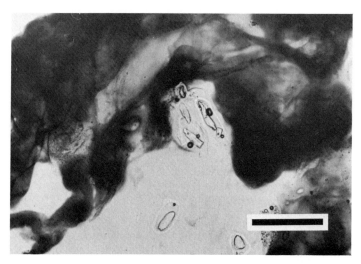

Figure 12.2: Vegetarian instant jelly powder containing carrageenan and guar gum stained with aqueous toluidine blue. Compare with mixture of same gums shown in *Figure 12.1*. Bar = 100 µm.

The use of more than one microtechnique is important when a mixture of gums is being examined. In addition to the results obtained by toluidine blue and iodine staining, further information can be gained by examination of the dry powder, that is the powder mounted in a non-aqueous mountant, such as light machine oil. This gives a quite different picture of the gum mixture. All gums show some birefringence when examined in the dry state, some being more birefringent than others. For example, the larger particles of xanthan gum show a full range of polarization colours and low, medium and high CMCs can be distinguished from one another by their appearance when viewed between crossed polars.

CMC shows its cellulose fibre origin most clearly when examined in a non-aqueous medium with brightfield illumination. When the polars are crossed, low viscosity CMC shows a full range of polarization colours, high viscosity CMC shows only white birefringence, whilst medium viscosity samples appear intermediate with only the thickest fibres showing any colour (*Figure 12.3*).

As with any identification work with the microscope, having a range of reference samples is useful as these can be combined to simulate an unknown mixture.

12.7 Identification of gums in food products

12.7.1 Dry products

These include powdered soups, sauce mixes, yoghurt stabilizers, powdered beverages and instant desserts.

The main difficulty with food products is that the gum may form only a small fraction of the total. However, this does not necessarily prevent its detection, especially if the microscopist can identify positively the other constituents present. Starch is a frequent ingredient of dry mixes and this in its various forms is easy to identify. Other constituents may need to be compared with reference materials, but a working knowledge of common ingredients is quickly built up.

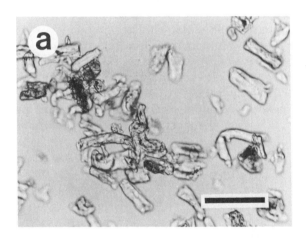

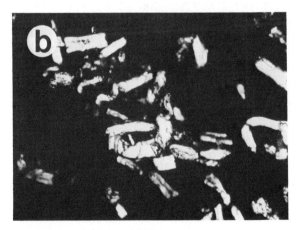

Figure 12.3: Food grade CMC mounted in oil. (a) Viewed with brightfield illumination. (b) Viewed between crossed polars. Bar = 100 μm.

Suggested approach.

1. Examine the dry powder mounted in light machine oil. This gives immediate positive results for starches and CMC. Palisade cells from guar and locust bean seed coats can often be seen at this stage.
2. Examine hydrated and acidified mounts after staining with toluidine blue. The seaweed gums and xanthan give the expected staining. Guar and locust bean may be understained but recognizable by their morphology.
3. Examine an iodine-stained mount: starches, tragacanth, seed gums, agar, unrefined carrageenan and Irish moss all give positive results.

12.7.2 Moist products

Moist foods which contain gums include sauces, confectionery jellies and pie fillings. In addition, most low-calorie foods contain gums, which act as thickeners giving body to the water which replaces oils and fats, for example low-calorie fat spreads, yoghurts and salad dressings.

Moist foods provide the microscopist with a challenge. The amount of food gum present is often low (< 2%) and it is usually well dispersed. Guar and carob gums are fairly easy to see because they retain their cellular structure and the presence of an anionic gum can often be confirmed, for example pectin in confectionery jellies (see Section 8.6.1). Starches, whether raw or gelatinized, alone or mixed with other gelling agents, are always relatively easy to demonstrate but individual gums can be difficult to identify if they are well dispersed.

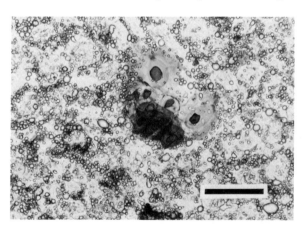

Figure 12.4: Smear of low-fat salad cream stained with aqueous toluidine blue showing guar cells. Bar = 100 μm

Suggested approach. Prepare a thin smear of the material. The product is already hydrated so avoid adding additional water unless this is really necessary.

Add 1 drop of 0.1% toluidine blue to the smear, leave for 1 min or longer if the product contains much fat. An alternative which is recommended for fat spreads is to place 1 drop of 0.1% toluidine blue on a coverglass, gently lower the smear on to this and, as soon as contact is made, invert the slide and leave for 1 min (*Figure 12.4*).

Aqueous iodine may be used to show the presence of starch but use iodine vapour to show its distribution in relation to other constituents (see Section 6.3.2).

Further reading

Blanshard JMV, Mitchell IR (eds) (1974) *Polysaccharides in Foods*. Butterworth, London.
Flint FO. (1990) Micro-technique for the identification of food hydrocolloids. *Analyst* **115,** 61–63.
Glicksman M. (1969) *Gum Technology in the Food Industry*. Academic Press, London.
Glicksman M. (ed.) (1982) *Food Hydrocolloids*, Vol. I. CRC Press, Boca Raton, FL.
Glicksman M. (ed.) (1983) *Food Hydrocolloids*, Vol. II. CRC Press, Boca Raton, FL.
Whistler RL, BeMiller JN. (eds.) (1993) *Industrial Gums Polysaccharides and their Derivatives*, 3rd edn. Academic Press, New York.

13 Food Emulsions

13.1 Introduction

Emulsions are two-phase systems containing droplets of one liquid dispersed in another, the droplets being of colloidal or microscopic size. One phase is oily and the other aqueous. The oil may be dispersed in water forming an oil-in-water (o/w) emulsion, or the aqueous droplets may be dispersed in oil giving a water-in-oil (w/o) emulsion.

Examples of o/w food emulsions are milk, cream, mayonnaise and salad cream. Butter, margarine and most low fat spreads are w/o emulsions (Dickinson, 1988). Food emulsions are rarely simple o/w or w/o emulsions. One or both phases may be partly solid; for example the water in ice cream is frozen and the fat in butter forms a network of tiny crystals surrounding the aqueous droplets. Emulsifiers are needed to make the emulsion and stabilizers may be added to act as thickeners of the aqueous phase. As well as dissolved ingredients, an emulsion may carry insolubles such as herbs or spices, e.g. powdered mustard in salad cream.

13.2 Microscopy of food emulsion constituents

Emulsions may be simple or difficult subjects to prepare for the microscope depending on what information about them is being sought. The first three experiments described show how easily some of the different ingredients present in an emulsion can be identified if the loss of emulsion structure is acceptable. These are followed by methods where the aim is to retain the fine structure of the emulsion whilst still demonstrating its constituents.

13.2.1 Solid fat in butter, margarines and low-fat spreads

Both butter and margarine contain fat crystals which are only 1–2 μm in size and the manufacturers of low-fat spreads aim to form fat crystals of a similar size. Crystals of this size may be difficult to see but, if the product is melted and allowed to cool, much larger fat crystals form which are visible with a 10× objective. To do this, place a small piece of the fatty product on a microscope slide and allow it to just melt on a hot plate or over hot water, apply a coverglass and allow to cool very slowly. View the large crystals which form with partly and fully crossed polars (*Figure 13.1*).

Figure 13.1: Spherulites of needle-shaped fat crystals formed when remelted low-fat spread cools. Unmounted preparation viewed between crossed polars. Bar = 100 μm.

13.2.2 Liquid fat and comparison of o/w and w/o emulsions

This is a suitable experiment to do with a low-calorie salad cream as the o/w emulsion and a low-fat spread as the w/o emulsion. Choose a fat spread that contains half the fat of butter (i.e. 38–40% fat).

Make a thin smear of each emulsion and stain for 5 min by the rapid Oil Red O method described in Section 7.3.2, but do not rinse. Quickly place a coverglass on the stained material, blot round the coverglass and examine. The oil droplets in the salad cream are coloured orange–red and the low-fat spread shows an orange–red matrix of fat containing colourless aqueous droplets. When viewed with crossed polars the preparation of the low-fat spread may show areas of birefringent fat (*Figures 13.2* and *13.3*).

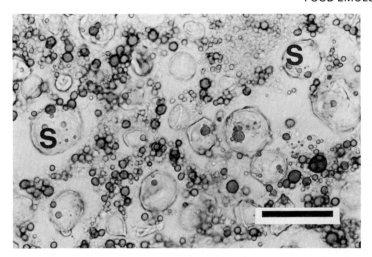

Figure 13.2: Smear of low-fat salad cream stained with Oil Red O. Note stained oil droplets, and unstained, cooked starch granules (S). Bar = 100 μm.

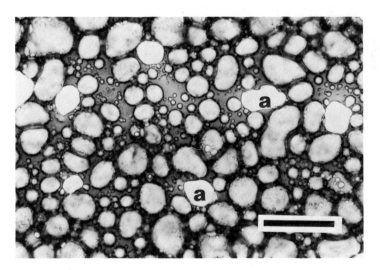

Figure 13.3: Smear of low-fat spread stained with Oil Red O. Note water droplets in stained fatty matrix and occasional air pockets (a). Bar = 100 μm.

13.2.3 Identification of the stabilizer in a low-calorie salad cream and a low-fat spread

Make two smears each of the low-calorie salad cream and low-fat spread used in Section 13.2.2. To one smear of each pair add 1 drop of aqueous 0.1% toluidine blue and to the other 1 drop of freshly diluted iodine, that is one part 1% iodine diluted to 20 ml with water. Leave for 1 min and add a coverglass to each.

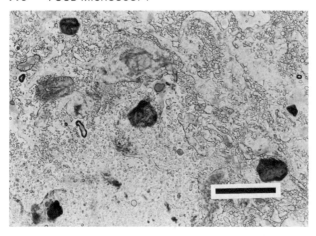

Figure 13.4: Smear of low-fat spread containing modified starch stained with aqueous iodine: note gelatinized starch granules. Bar = 100 µm.

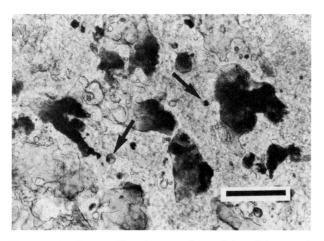

Figure 13.5: Smear of low-fat spread containing alginate stained with aqueous toluidine blue. Some alginate has remained in droplets (arrows) but much has been released by smearing. Bar = 100 µm.

The iodine will show the presence of starch in the salad cream and possibly in the low fat spread (*Figure 13.4*).

The toluidine blue will stain alginates or other gums present a purple colour (see *Figure 13.5,* and *Figure 12.4* in Chapter 12).

Any plant tissues present in the salad cream, such as ground mustard seed, will be stained by the toluidine blue, the mucilage cells of mustard stains a pink colour and the lignified seed coat turquoise.

13.3 Emulsion microstructure

The experiments described in the previous section will almost certainly alter its structure. Where the aim is a study of emulsion structure and the relationship of its constituents to one another, greater care is needed and the methods of demonstrating constituents must be modified.

Although oil-soluble colours show the presence of liquid fat in emulsions, the aqueous solvents involved tend to disrupt the emulsion structure, particularly that of o/w emulsions. Vapour staining of emulsion smears using osmium tetroxide is recommended but only if the staining can be done safely as described in Chapter 7 (*Figure 13.6*).

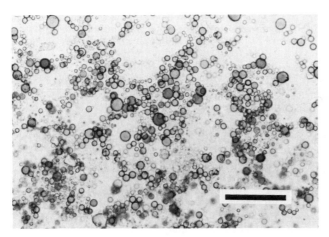

Figure 13.6: Smear of low-fat salad cream stained with osmium tetroxide vapour. The stained oil globules have a similar size range to those stained with Oil Red O (*Figure 13.2*). This is an indication of the emulsion's stability. Bar = 100 μm.

Cooked modified starch and dextrins from starch are used widely as thickeners in low-fat spreads and low-calorie salad creams. These can be demonstrated without dispersing the starch or dextrin by staining emulsion smears with iodine vapour as described in Chapter 6 (*Figures 13.7* and *13.8*).

For the demonstration of food gums in w/o emulsions, place one very small drop of 0.1% toluidine blue on a coverglass, next prepare the emulsion smear and immediately lower it on to the coverglass. As soon as contact is made invert the slide and leave for 1 min before examination. The aqueous droplets in the emulsion take up the stain, giving a picture very similar to that of *Figure 13.7*.

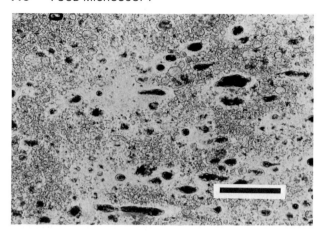

Figure 13.7: Smear of low-fat spread containing modified starch stained with iodine vapour. Although smearing has distorted some of the water droplets containing starch, iodine vapour gives a truer picture of the starch distribution than that seen in *Figure 13.4*. Bar = 100 µm.

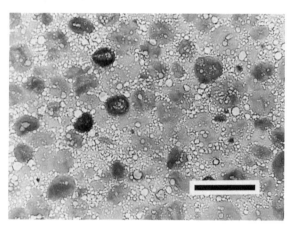

Figure 13.8: Smear of low-fat salad cream stained with iodine vapour showing gelatinized starch granules and unstained oil globules. Compare distribution and size of starch granules with those seen in *Figure 13.2* where the starch was free to move and swell. Bar = 100 µm.

Further reading

Brooker BE. (1979) Milk and its products. In *Food Microscopy* (ed. JG Vaughan). Academic Press, London, pp. 273–308.

Dickinson E, Stainsby G. (1982) *Colloids in Food*. Applied Science, London.

Dickinson E. (1988) The structure and stability of emulsions. In *Food Structure: its Creation and Evaluation* (eds JMV Blanshard, JR Mitchell). Butterworths, London, pp. 41–57.

Flint FO. (1984) Applications of light microscopy in food analysis. *Microscope* **32**, 133–140.

Mageean P, Jones S. (1989) Low-fat spread products. *Food Sci. Technol. Today* **3**, 162–164.

Appendix

Equipment

Equipment used in the experimental methods includes the following.

Cryostat, accessories and knife sharpening service: Bright Instrument Company Ltd, St Margarets Way, Stukeley Meadows Road, Huntingdon, Cambridgeshire PE18 6EB, UK.

Forceps, cutting needles, section lifter and other small tools: Raymond A. Lamb, 6 Sunbeam Road, London NW10 6JL, UK.

Sheet polaroid material 75×75 mm and 100×100 mm: Hargreaves Photographic Ltd, Unit 3, 214 Purley Way, Croydon, London CR0 4XG, UK. (Also branches in Leeds, Liverpool and Manchester.)

Howard mould count cell and coverglass: Graticules Ltd, Morley Road, Tonbridge, Kent TN9 1RN, UK.

Materials

Cryostat embedding medium: Tissue-Tek OCT, Raymond A. Lamb, 6 Sunbeam Road, London NW10 6JL, UK.

Starches, gums and Biological Commission Certified stains: Sigma Chemical Company Ltd, Fancy Road, Poole, Dorset BH17 7NH, UK. US address: Sigma Chemical Co., 3050 Spruce St, St Louis, MO 63178.

Osmium tetroxide in sealed ampoules: Taab Laboratories Equipment Ltd, Unit 3 Minerva House, Calleva Industrial Park, Aldermaston, Reading, Berkshire RG7 4QW, UK.

Stains: Aldrich Chemical Co., The Old Brickyard, New Road, Gillingham, Dorset SP8 4JL, UK. US address: Aldrich Chemical Co., Inc. 1001 W. St Paul Ave., PO Box 355, Milwaukee, WI 53201.

Index

In Situ Hybridization

A.R. Leitch, T. Schwarzacher, D. Jackson & I.J. Leitch
respectively Queen Mary and Westfield College, London, UK; John Innes Research Centre, Norwich, UK; USDA, Plant Gene Expression Center, Albany, California, USA; and Royal Botanic Gardens, Kew, UK

In situ hybridization is a powerful link beween cellular and molecular biology. This practical guide provides a comprehensive description of *in situ* hybridization, from background information to detailed methodology and practical applications. The book's clarity of approach and up-to-date coverage of methods and troubleshooting makes it the ideal introduction for all first-time users and a valuable companion for experienced researchers.

Contents

Introduction; Nucleic acid sequences located *in situ*; The material; Nucleic acid probes, labels and labelling methods; Denaturation, hybridization and washing; Detection of the *in situ* hybridization sites; Imaging systems and the analysis of signal; The *in situ* hybridization schedule (including troubleshooting); The future of *in situ* hybridization. Appendix: Suppliers of reagents and *in situ* hybridization kits; Buffers.

Of interest to:

Advanced undergraduates and postgraduate students of molecular biology, cell biology and genetics.

Paperback; 128 pages; 1-872748-48-1; 1994

Biological Microtechnique

J. Sanderson
Sir William Dunn School of Pathology, Oxford, UK

Although many significant advances have been made in biological specimen preparation during the past 20 years no new practical guide to the techniques has been published in this time. As a result of the recent resurgence of interest in light microscopy, particularly confocal techniques, this new, up-to-date book will benefit both novices and experienced microscopists seeking to extend their repertoire of techniques. A poorly-prepared specimen inevitably leads to unreliable results. This new book therefore describes both new and classical methods of slide-making in an easy-to-read, easy-to-understand format. It contains a wealth of practical detail which will provide a firm grounding in preparative methods for light microscopy.

Contents

Fixation; Tissue processing; Microtomy; Other preparative methods; Staining and dyeing; Finishing the preparation. Appendices: safety; removing dyestains; restaining faded sections; restoring tissues; cleaning glassware; physiological solutions; saturation.

Of interest to:

Junior researchers, laboratory technicians, skilled amateur microscopists, undergraduates, school science teachers and students.

Paperback; 240 pages; 1-872748-42-2; 1994

ALSO AVAILABLE FROM BIOS SCIENTIFIC PUBLISHERS LTD

Flow Cytometry

M.G. Ormerod
Scientific Consultant, Reigate, Surrey, UK

Flow cytometry is a specialised form of microscopy for measuring the
properties of single cells. Flow cytometers are becoming widely used in
clinical and research laboratories and there is an increasing need for
non-specialists to have an understanding of this technology. *Flow Cytometry*
is a practical guide to the instrumentation and the application of this method
in mammalian cell biology. The book is easy to read and contains sufficient
information and references to allow the reader to pursue any particular
application in greater depth. It covers the routine applications of flow
cytometry and introduces the reader to the more recent applications of this
exciting technology.

Contents

What is flow cytometry; Instrumentation; Fluorescence;
Immunofluorescence; Analysis of DNA; Study of cell proliferation and death;
Other applications. *Appendices:* Glossary; Suppliers; Learned societies.

Of interest to:

Postgraduates, clinicians, researchers, technicians and all first-time users.

Paperback; 88 pages;   1-872748-39-2; 1994

Scientific PhotoMACROgraphy

B. Bracegirdle
Cheltenham, UK

This book provides authoritative advice on recording specimens at magnifications between x1 and x50. It covers choice of lenses, cameras and accessories and gives valuable tips on making up special parts to suit particular purposes. The different approaches required for transmitted-light and for reflected-light work are also described and details given of the many ways of recording the images produced. Extensive reference data are provided in readily-accessible Nomograms and tables and some 40 half-tones show how the equipment should be set up. *Scientific PhotoMACROgraphy* minimizes the frustration often experienced in this difficult area of work and is an invaluable practical guide to making high quality records consistently.

Contents

The scope of the process; Obtaining the magnification; Working with transmitted light; Working with reflected light; General remarks on illumination and exposure; Estimating exposure in macro-range photography; Recording the image.

Of interest to:

2nd year undergraduates and above; any researcher using the microscope at low resolution.

Paperback; 120 pages; 1-872748-49-X; 1994

ORDERING DETAILS

Main address for orders

BIOS Scientific Publishers Ltd
St Thomas House, Becket Street,
Oxford OX1 1SJ, UK
Tel: +44 1865 726286
Fax: +44 1865 246823

Australia and New Zealand
DA Information Services
648 Whitehorse Road, Mitcham, Victoria 3132, Australia
Tel: (03) 873 4411
Fax: (03) 873 5679

India
Viva Books Private Ltd
4346/4C Ansari Road, New Delhi 110 002, India
Tel: 11 3283121
Fax: 11 3267224

Singapore and South East Asia
(Brunei, Hong Kong, Indonesia, Korea, Malaysia, the Philippines,
Singapore, Taiwan, and Thailand)
Toppan Company (S) PTE Ltd
38 Liu Fang Road, Jurong, Singapore 2262
Tel: (265) 6666
Fax: (261) 7875

USA and Canada
Books International Inc
PO Box 605, Herndon, VA 22070, USA
Tel: (703) 435 7064
Fax: (703) 689 0660

Payment can be made by cheque or credit card (Visa/Mastercard, quoting number and expiry date). Alternatively, a *pro forma* invoice can be sent.

Prepaid orders must include £2.50/US$5.00 to cover postage and packing for one item and £1.25/US$2.50 for each additional item.